Annals of Mathematics Studies

Number 61

SINGULAR POINTS OF COMPLEX HYPERSURFACES

BY

John Milnor

PRINCETON UNIVERSITY PRESS

AND THE

UNIVERSITY OF TOKYO PRESS

PRINCETON, NEW JERSEY

1968

Preface

The topology associated with a singular point of a complex curve has fascinated a number of geometers, ever since K. BRAUNER* showed in 1928 that each such singular point can be described in terms of an associated knotted curve in the 3-sphere. Recently E. BRIESKORN has brought new interest to the subject by discovering similar examples in higher dimensions, thus relating algebraic geometry to higher dimensional knot theory and the study of exotic spheres.

This manuscript will study singular points of complex hypersurfaces by introducing a fibration which is associated with each singular point.

As prerequisites the reader should have some knowledge of basic algebra and topology, as presented for example in LANG, *Algebra* or VAN DER WAERDEN, *Modern Algebra*, and in SPANIER, *Algebraic Topology*.

I want to thank E. Brieskorn, W. Casselman, H. Hironaka, and J. Nash for helpful discussions; and E. Turner for preparing notes on an earlier version of this material. Also I want to thank the National Science Foundation for support. Work on this manuscript was carried out at Princeton University, the Institute for Advanced Study, The University of California at Los Angeles, and the University of Nevada.

---

* See the Bibliography. Proper names in capital letters will always indicate a reference to the Bibliography.

CONTENTS

Annals of Mathematics Studies

Number 61

§1. INTRODUCTION

Let $f(z_1, \ldots, z_{n+1})$ be a non-constant polynomial in $n+1$ complex variables, and let V be the algebraic set consisting of all $(n+1)$-tuples

$$\mathbf{z} = (z_1, \ldots, z_{n+1})$$

of complex numbers with $f(\mathbf{z}) = 0$. (Such a set is called a *complex hypersurface*.) We want to study the topology of V in the neighborhood of some point $\mathbf{z}^0$.

We will use the following construction, due to BRAUNER. Intersect the hypersurface V with a small sphere S_ε centered at the given point $\mathbf{z}^0$. Then the topology of V within the disk bounded by S_ε is closely related to the topology of the set

$$K = V \cap S_\varepsilon \ .$$

(Compare §2.10 and §2.11.)

As an example, if $\mathbf{z}^0$ is a *regular point* of f (that is if some partial derivative $\partial f/\partial z_j$ does not vanish at $\mathbf{z}^0$) then V is a smooth manifold of real dimension $2n$ near $\mathbf{z}^0$. The intersection K is then a smooth $(2n-1)$-dimensional manifold, diffeomorphic to the $(2n-1)$-sphere, and K is embedded in an unknotted manner in the $(2n+1)$-sphere S_ε. (See §2.12.)

By way of contrast, consider the polynomial

$$f(z_1, z_2) = z_1^p + z_2^q$$

in two variables, with a *critical point* $(\partial f/\partial z_1 = \partial f/\partial z_2 = 0)$ at the origin. Assume that the integers p, q are relatively prime and ≥ 2.

ASSERTION (Brauner). *The intersection of* $V = f^{-1}(0)$ *with a sphere* S_ε *centered at the origin is a knotted circle of the type known as a "torus knot of type* (p, q)*" in the 3-sphere* S_ε.

[*Proof:* It is easily verified that the intersection K lies in the torus consisting of all (z_1, z_2) with $|z_1| = \xi$, $|z_2| = \eta$ where ξ and η are positive constants. In fact, K consists of all pairs $(\xi e^{qi\theta}, \eta e^{pi\theta + \pi i/q})$ as the parameter θ ranges from 0 to 2π: Thus K sweeps around the torus q times in one coordinate direction and p times in the other.]

For example the torus knot of type (2, 3) is illustrated in Figure 1.

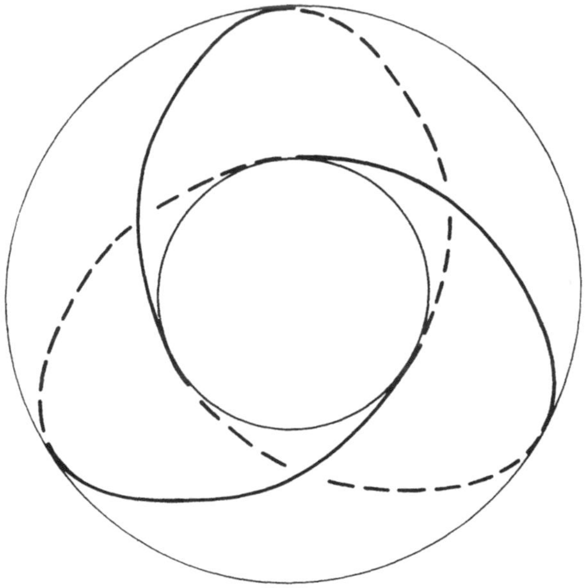

Figure 1. The torus knot of type (2, 3).

(By using more complicated polynomials one can of course arrive at much more complicated knots. Compare §10.11.)

BRIESKORN has studied higher dimensional analogues of these torus knots. For example let $V(3, 2, 2, \ldots, 2)$ be the locus of zeros of the polynomial

$$f(z_1, \ldots, z_{n+1}) = z_1^{\,3} + z_2^{\,2} + \cdots + z_{n+1}^{\,2} .$$

For all odd values of n this hypersurface intersects S_ε in a smooth manifold K which is homeomorphic to the sphere S^{2n-1}. In some cases (for example when $n = 3$) K is diffeomorphic to the standard $(2n-1)$-sphere, while in other cases (for example $n = 5$) K is an "exotic" sphere. But in all cases K is embedded in a knotted manner in the $(2n+1)$-sphere S_ε.

These Brieskorn spheres will be studied in more detail in §9.

The object of this paper is to introduce a fibration which is useful in describing the topology of such intersections

$$K = V \cap S_\varepsilon \subset S_\varepsilon \quad .$$

Here are some of the main results, which will be proved in Sections 4 through 7.

FIBRATION THEOREM. *If* $\mathbf{z}^0$ *is any point of the complex hypersurface* $V = f^{-1}(0)$ *and if* S_ε *is a sufficiently small sphere centered at* $\mathbf{z}^0$, *then the mapping*

$$\phi(\mathbf{z}) = f(\mathbf{z})/|f(\mathbf{z})|$$

from $S_\varepsilon - K$ *to the unit circle is the projection map of a smooth fiber bundle**. *Each fiber*

$$F_\theta = \phi^{-1}(e^{i\theta}) \subset S_\varepsilon - K$$

is a smooth parallelizable 2n-*dimensional manifold.*

If the polynomial f has no critical points near $\mathbf{z}^0$, except for $\mathbf{z}^0$ itself, then we can give a much more precise description.

THEOREM. *If* $\mathbf{z}^0$ *is an isolated critical point of* f, *then each fiber* F_θ *has the homotopy type of a bouquet* $S^n \vee \cdots \vee S^n$ *of* n-*spheres, the number of spheres in this bouquet (i.e., the middle Betti number of* F_θ*), being strictly positive. Each fiber can be considered as the interior of a smooth compact manifold-with-boundary,*

$$\text{Closure}\,(F_\theta) = F_\theta \cup K\ ,$$

* The term "fiber bundle" will be used as a synonym for "locally trivial fiber space."

where the common boundary K *is an* $(n-2)$*-connected manifold.*

Thus all of the fibers F_θ fit around their common boundary K in the manner illustrated in Figure 2. The smooth manifold K is connected if $n \geq 2$, and simply connected if $n \geq 3$.

Here is a more detailed outline of what follows. Section 2 describes elementary properties of real algebraic sets, following WHITNEY. A fundamental lemma concerning the existence of real analytic curves on real algebraic sets is proved in §3. All of the subsequent proofs rely on this lemma. The basic fibration theorem is proved in §4. Further details on the topology of K and F_θ are obtained in §5.

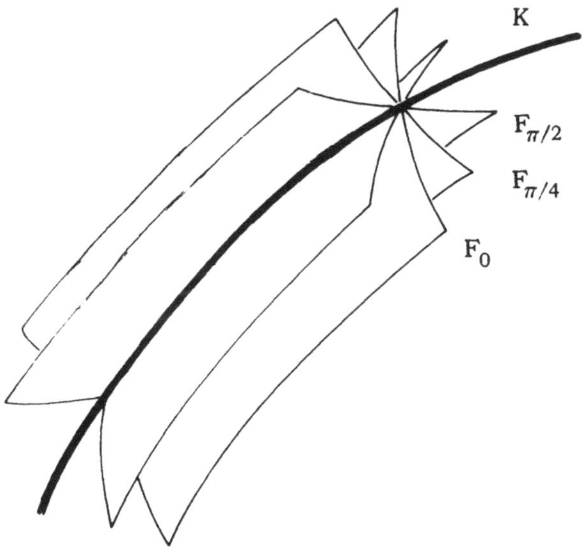

Figure 2.

Next we introduce the additional hypothesis that the origin is an *isolated* critical point of f. Then a much more precise description of the fiber is possible (§6), and a precise formula for the middle Betti number of the fiber is given (§7). The topology of the intersection K is then described in terms of a certain polynomial $\Delta(t)$ with integer coefficients which generalizes the Alexander polynomial of a knot. (§8.)

The Brieskorn examples of singular varieties which are topologically manifolds are described in §9, and the classical theory of singular points of complex curves is described in §10. The last section proves a generalization of the fibration theorem to certain systems of real polynomials. As an example, a polynomial description of the Hopf fibrations is given.

Two appendices conclude the presentation.

§2. ELEMENTARY FACTS ABOUT REAL OR COMPLEX ALGEBRAIC SETS

Let Φ be any infinite field, and let Φ^m be the coordinate space consisting of all m-tuples $x = (x_1, \ldots, x_m)$ of elements of Φ. (We are principally interested in the case where Φ is the field R of real numbers or the field C of complex numbers.)

DEFINITION. A subset $V \subset \Phi^m$ is called an *algebraic set** if V is the locus of common zeros of some collection of polynomial functions on Φ^m.

The ring of all polynomial functions from Φ^m to Φ will be denoted by the conventional symbol $\Phi[x_1, \ldots, x_m]$. Let

$$I(V) \subset \Phi[x_1, \ldots, x_m]$$

be the ideal consisting of those polynomials which vanish throughout V. The Hilbert "basis" theorem asserts that every ideal is spanned (as $\phi[x_1, \ldots, x_m]$-module) by some finite collection of polynomials. It follows that every algebraic set V can be defined by some finite collection of polynomial equations.

An important consequence of the Hilbert basis theorem is the following:

2.1 *Descending chain condition.* Any nested sequence $V_1 \supset V_2 \supset V_3 \supset \cdots$ of algebraic sets must terminate or stabilize $(V_i = V_{i+1} = V_{i+2} = \cdots)$ after a finite number of steps.

* It is customary in algebraic geometry to allow as "points" of V also m-tuples of elements belonging to some fixed algebraically closed extension field of Φ; but I do not want to allow this..

Note that the union $V \cup V'$ of any two algebraic sets V and V' in Φ^m is again an algebraic set.

A non-vacuous algebraic set V is called a *variety* or an *irreducible algebraic set* if it cannot be expressed as the union of two proper algebraic subsets. Note that V is irreducible if and only if $I(V)$ is a prime ideal. If V is irreducible, then the field of quotients f/g with f and g in the integral domain

$$\Phi[x_1, \ldots, x_m]/I(V)$$

is called the *field of rational functions* on V. Its transcendence degree over Φ is called the algebraic *dimension* of V over Φ.

If W is a proper subvariety of V, note that the dimension of W is less than the dimension of V. (See for example LANG, *Algebraic Geometry*, p. 29.)

Now let $V \subset \Phi^m$ be any non-vacuous algebraic set. Choose finitely many polynomials $f_1, \ldots, f_k$ which span the ideal $I(V)$ and, for each $\mathbf{x} \in V$, consider the $k \times m$ matrix $(\partial f_i/\partial x_j)$ evaluated at $\mathbf{x}$. Let ρ be the largest rank which this matrix attains at any point of V.

DEFINITION. A point $\mathbf{x} \in V$ is called *non-singular* or *simple* if the matrix $(\partial f_i/\partial x_j)$ attains its maximal rank ρ at $\mathbf{x}$; and *singular** if

$$\text{rank}\ (\partial f_i(\mathbf{x})/\partial x_j) < \rho\ .$$

Note that this definition does not depend on the choice of $\{f_1, \ldots, f_k\}$. (For if we add an extra polynomial $f_{k+1} = g_1 f_1 + \cdots + g_k f_k$ the resulting new row in our matrix will be a linear combination of the old rows.)

LEMMA 2.2. *The set $\Sigma(V)$ of all singular points of V forms a proper algebraic subset (possibly vacuous) of V.*

* This definition is certainly the correct one whenever V is a variety, or a union of varieties all of which have the same dimension. In other cases it does not correspond too well to intuitive expectations. For example if V is the union of a point and a line, then only the point is non-singular.

For a point x of V belongs to $\Sigma(V)$ if and only if every $\rho \times \rho$ minor determinant of $(\partial f_i/\partial x_j)$ vanishes at x. Thus $\Sigma(V)$ is determined by polynomial equations.

Now let us specialize to the case of a real or complex algebraic set.

THEOREM 2.3 (Whitney). *If Φ is the field of real (or complex) numbers, then the set $V - \Sigma(V)$ of non-singular points of V forms a smooth, non-vacuous manifold. In fact this manifold is real (or complex) analytic, and has dimension $m - \rho$ over Φ.*

The reader is referred to WHITNEY, *Elementary Structure of Real Algebraic Varieties*, for the elegant proof of 2.3.

In the case of an irreducible V, Whitney shows that *the dimension of the analytic manifold $V - \Sigma(V)$ over Φ is precisely equal to the algebraic dimension of V over Φ.*

Here is another basic result.

THEOREM 2.4 (Whitney). *For any pair $V \supset W$ of algebraic sets in a real or complex coordinate space, the difference $V - W$ has at most a finite number of topological components.*

For example, V itself has only finitely many components; and the smooth manifold $V - \Sigma(V)$ has only finitely many components.

A proof of 2.4, only slightly different from WHITNEY'S proof, will be given in Appendix A.

Here are three examples. (Compare Figure 3.) Each example will be a curve in the real plane having the origin as unique singular point.

EXAMPLE A. The variety consisting of all (x, y) in R^2 with

$$y^2 - x^2(1 - x^2) = 0$$

illustrates the most well behaved and easily understood type of singular point, a "double point" at which two real analytic branches with distinct tangents (namely $y = x\sqrt{1-x^2}$ and $y = -x\sqrt{1-x^2}$) cross each other.*

* This can also be seen from the parametric representation $x = \sin\theta$, $2y = \sin 2\theta$ (which shows that the curve is a "Lissajous figure").

(Figure 3-A. For a definition of the term "branch" see §3.3.)

EXAMPLE B. The cubic curve

$$y^2 - x^2(x-1) = 0$$

of Figure 3-B has an isolated point at the origin; yet this curve is also irreducible.

(REMARK. Over the field of complex numbers, examples of this type cannot occur. In fact a theorem of RITT implies that the manifold of simple points of a complex variety V is everywhere dense in V. Compare VAN DER WAERDEN *Zur algebraische Geometrie III*, or *Algebraische Geometrie*, p. 134.)

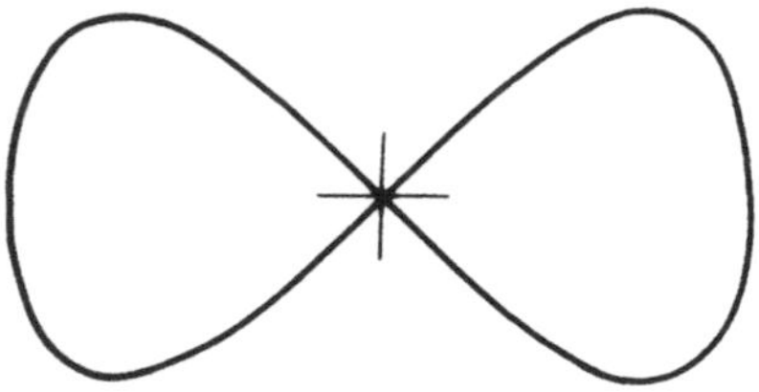

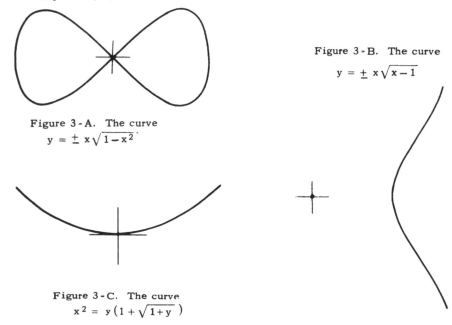

Figure 3-A. The curve $y = \pm x\sqrt{1-x^2}$

Figure 3-B. The curve $y = \pm x\sqrt{x-1}$

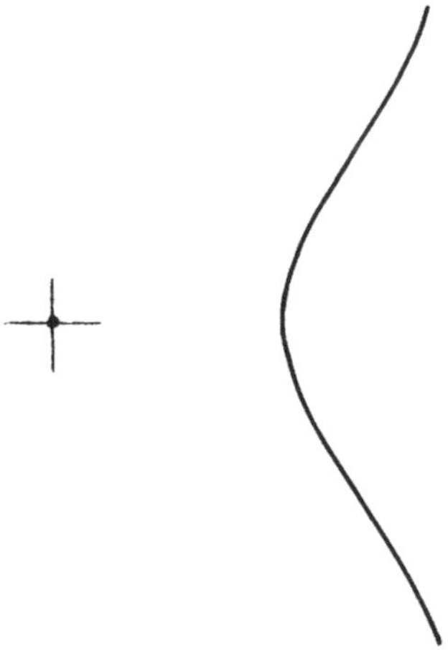

Figure 3-C. The curve $x^2 = y(1+\sqrt{1+y})$

EXAMPLE C. The equation $y^3 = x^{100}$ can be solved for y as a 33-times differentiable function of x, yet this equation defines a variety $V \subset R^2$ which has a singular point at the origin. The equation $y^3 + 2x^2y - x^4 = 0$, which is illustrated in Figure 3-C, can actually be solved for y as a real analytic function* of x, but this equation also defines a variety having a singular point at the origin.

If we allow x and y to vary over the complex numbers, then the phenomenon becomes easier to understand. In fact, the complex curve $y^3 = x^{100}$ is "knotted" near the origin (compare §1), and the complex curve $y^2 + 2x^2y = x^4$ has three distinct non-singular branches passing through the origin.

REMARK. A complex variety can never be a smooth manifold throughout a neighborhood of a singular point.

Proof: Suppose that the complex variety V were a differentiable manifold of class $\mathcal{C}^1$ throughout a neighborhood U of the origin in C^m. The tangent space of this smooth manifold $U \cap V$ at any simple point is clearly a vector space over the complex numbers. Since the simple points are dense (see the Remark above), it follows by continuity that the (real) tangent space $T_{\mathbf{z}} \subset C^m$ of $U \cap V$ at an arbitrary point $\mathbf{z}$ is actually a complex vector space. (That is, $T_{\mathbf{z}} = iT_{\mathbf{z}}$.) Now replacing U by a smaller neighborhood U', and renumbering the coordinates if necessary, the implicit function theorem shows that $U' \cap V$ can be considered as the graph of a $\mathcal{C}^1$-smooth mapping F from an open subset of the $(z_1, \ldots, z_n)$ coordinate space into the $(z_{n+1}, \ldots, z_m)$ coordinate space. The derivative of F at each point is complex linear, hence the Cauchy-Riemann equations are satisfied, and F is complex analytic. This proves that $U' \cap V$ is a complex manifold. Next let $h(\mathbf{z})$ be any complex analytic function, defined in a neighborhood of $\mathbf{0}$, which vanishes on V, and let $f_1, \ldots, f_k$ be polynomials which span the prime ideal $I(V) \subset C[z_1, \ldots, z_m]$. The local analytic Nullstellensatz (see for example, GUNNING and ROSSI, p. 90) asserts that some power h^s can be expressed as a linear combination $a_1f_1 + \cdots + a_kf_k$ where $a_1, \ldots, a_k$ are germs of analytic functions. Passing to the larger ring $C[[\mathbf{z}]]$ consisting of all formal power series at the origin, it follows a fortiori that h^s belongs to the corresponding ideal $C[[\mathbf{z}]]\, I(V)$. But this

* *Proof:* Solve for x^2 as a function $x^2 = \phi(y) = y + y\sqrt{1+y}$, and note that ϕ^{-1} is defined and analytic throughout the interval $[0, \infty)$.

ideal can be expressed as an intersection of prime ideals (see LEFSCHETZ, *Algebraic Geometry,* p. 91). Therefore h itself must belong to the ideal $C[[\mathbf{z}]]\,I(V)$; which is spanned by $f_1, \ldots, f_k$ in $C[[\mathbf{z}]]$. Taking derivatives, this implies that the covector $dh(\mathbf{0})$ can be expressed as a complex linear combination of the covectors $df_1(\mathbf{0}), \ldots, df_k(\mathbf{0})$. Now it follows easily that the matrix $(\partial f_i/\partial z_j)$ has rank $m-n$ at $\mathbf{0}$; which proves that the origin cannot be a singular point of V.

The reader should have no difficulty in checking that Examples A, B and C do have the properties ascribed to them, making use of the following.

LEMMA 2.5. *Let* V *be the real or complex algebraic set defined by a single polynomial equation* $f(x) = 0$; *with* f *irreducible. In the real case make the additional hypothesis that* V *contains a regular point** *of* f. *Then every polynomial which vanishes on* V *is a multiple of* f.

Hence V is irreducible, and the singular set $\Sigma(V)$ is precisely the intersection of V with the set of critical points of f.

Proof: In the complex case this follows immediately from the Hilbert Nullstellensatz. In the real case, express $V \subset R^m$ as a union $V_1 \cup \cdots \cup V_k$ of varieties. Since a neighborhood of the regular point in V is an $(m-1)$-dimensional manifold, a topological argument shows that at least one of the V_j must have dimension $m-1$. Hence, according to Whitney, the quotient domain $R[x_1, \ldots, x_m]/I(V_j)$ has transcendence degree $m-1$ over R. But the quotient of $R[x_1, \ldots, x_m]$ by the principal prime ideal (f) clearly also has transcendence degree $m-1$ over R. Since

$$(f) \subset I(V_j),$$

a standard argument shows that $(f) = I(V_j)$. (See for example LANG, *Introduction to Algebraic Geometry,* p. 29.)

Therefore the set of zeros of f coincides with the variety V_j. This proves that $V = V_j$, and hence that $I(V)$ is equal to the principal ideal (f).

---

* This hypothesis is needed to avoid examples such as $x^2 + y^2 + z^2 = 0$.

(REMARK. The analogous statement is true over any locally compact field, but not over an arbitrary field. For example the irreducible polynomial $x^2 - y - y^3$ over the field of rational numbers has no critical points in the rational plane, and has just one rational zero.)

Now let us draw further consequences from Whitney's two theorems.

COROLLARY 2.6. *Any real or complex algebraic set* V *can be expressed as a finite disjoint union*

$$V = M_1 \cup M_2 \cup \cdots \cup M_p ,$$

where each M_j *is a smooth manifold with only finitely many components. Similarly any difference* $V - W$ *of varieties can be expressed as such a finite union.*

Proof: Let $M_1 = V - \Sigma(V)$ be the set of simple points of V, let $M_2 = \Sigma(V) - \Sigma(\Sigma(V))$ be the set of simple points of $\Sigma(V)$, and so on. This construction must stop after finitely many stages, since the sequence

$$V \supset \Sigma(V) \supset \Sigma(\Sigma(V)) \supset \cdots$$

must terminate by 2.1. Clearly V is the disjoint union of the manifolds M_i.

Similarly $V - W$ can be expressed as the disjoint union $M_1' \cup M_2' \cup \cdots \cup M_p'$ where each

$$M_i' = M_i - (W \cap M_i)$$

is a smooth manifold with finitely many components by 2.4.

The following lemma is frequently useful. As usual let Φ denote the real or complex numbers.

Let $M_1 = V - \Sigma(V)$ be the manifold of simple points of an algebraic set $V \subset \Phi^m$, and let g be a polynomial function on Φ^m.

LEMMA 2.7. *The set of critical points* of the restricted function* $g|M_1$ *from* M_1 *to* Φ *is equal to the intersection of* M_1 *with the algebraic set* W *consisting of all points* $x \in V$ *at which the matrix*

* A *critical point* of a smooth mapping between smooth manifolds is a point of the first manifold at which the induced linear mapping between tangent spaces fails to be surjective.

$$\begin{bmatrix} \partial g/\partial x_1 & \cdots & \partial g/\partial x_m \\ \partial f_1/\partial x_1 & \cdots & \partial f_1/\partial x_m \\ \vdots & & \vdots \\ \partial f_k/\partial x_1 & \cdots & \partial f_k/\partial x_m \end{bmatrix}$$

has rank $\leq \rho$; *where* $f_1, \ldots, f_k$ *denote polynomials spanning* $I(V)$.

Proof: Near any point of M_1 we can choose a (real or complex) analytic system of local coordinates $u_1, \ldots, u_m$ for Φ^m so that M_1 corresponds to the locus $u_1 = \cdots = u_\rho = 0$. Then $u_{\rho+1}, \ldots, u_m$ can be taken as local coordinates on M_1. Note that $\partial f_i/\partial u_j$, evaluated at a point of M_1, is zero for $j \geq \rho + 1$ (since f_i vanishes on M_1). Since the matrix $(\partial f_i/\partial u_j)$ is column equivalent to the matrix $(\partial f_i/\partial x_\ell)$ and therefore has rank ρ, it follows that the first ρ columns of $(\partial f_i/\partial u_j)$ must be linearly independent.

Now the enlarged matrix

$$\begin{bmatrix} \partial g/\partial u_1 & \cdots & \partial g/\partial u_m \\ \partial f_1/\partial u_1 & \cdots & \partial f_1/\partial u_m \\ \vdots & & \vdots \\ \partial f_k/\partial u_1 & \cdots & \partial f_k/\partial u_m \end{bmatrix}$$

will have the same rank ρ if and only if

$$\partial g/\partial u_{\rho+1} = \cdots = \partial g/\partial u_m = 0 ;$$

or in other words if and only if the given point is a critical point of $g | M_1$. Since this new matrix is column equivalent to the matrix given in 2.7, this completes the proof.

COROLLARY 2.8. *A polynomial function* g *on* $M_1 = V - \Sigma(V)$ *can have at most a finite number of critical values.*

(A *critical value* $g(x) \in \Phi$ is the image under g of a critical point.)

Proof: The set of critical points of $g | M_1$ can be expressed as a difference $W - \Sigma(V)$ of algebraic sets, and hence can be expressed as a finite union of smooth manifolds,

$$W - \Sigma(V) = M_1' \cup \cdots \cup M_p' ,$$

where each M_i' has only finitely many components.

Each point $x \in M_i'$ is a critical point of the smooth function $g \mid M_1$, so a fortiori it is a critical point of the restricted function $g \mid M_i'$. Since all points of M_i' are critical, it follows that g is constant on each component of M_i'. Therefore the image $g(M_i')$ is a finite set. But the union

$$g(M_1') \cup \cdots \cup g(M_p')$$

is precisely the set of all critical values of $g \mid M_1$. This completes the proof.

Again let V be any real or complex algebraic set. Let x^0 be either a simple point of V or an isolated point of the singular set $\Sigma(V)$.

COROLLARY 2.9. *Every sufficiently small sphere S_ε centered at x^0 intersects V in a smooth manifold (possibly vacuous).* *

Proof: In the real case this follows by applying 2.8 to the polynomial function

$$r(x) = \|x - x^0\|^2 .$$

If ε^2 is smaller than any positive critical value of $r \mid (V - \Sigma(V))$ then ε^2 will be a regular value, hence its inverse image

$$r^{-1}(\varepsilon^2) \cap (V - \Sigma(V)) = S_\varepsilon \cap (V - \Sigma(V))$$

will be a smooth manifold K. But if ε is small enough then S_ε will not meet $\Sigma(V)$, hence K will equal $S_\varepsilon \cap V$.

The corresponding statement in the complex case follows immediately, since every complex variety in C^m can be thought of as a real variety in R^{2m}.

Let D_ε denote the closed disk consisting of all x with $\|x - x^0\| \leq \varepsilon$. Again let x^0 be either a simple point or an isolated singular point of V.

* The proof will actually show that the intersection is *transverse*: that is every vector based at a point of $V \cap S_\varepsilon$ can be expressed as the sum of a vector tangent to V and a vector tangent to S_ε.

THEOREM 2.10. *For small* ε *the intersection of* V *with* D_ε *is homeomorphic to the cone over* $K = V \cap S_\varepsilon$. *In fact the pair* $(D_\varepsilon, V \cap D_\varepsilon)$ *is homeomorphic to the pair consisting of the cone over* S_ε *and the cone over* K.

Here, by the *cone* over K, denoted Cone (K), we mean the union of all line segments

$$tk + (1-t)x^0, \qquad 0 \le t \le 1,$$

joining points $k \in K$ to the base point x^0. The set Cone (S_ε), defined similarly, is of course precisely equal to D_ε.

Thus if we can identify the manifold K, and the way in which K is embedded in S_ε, then we will have completely determined the topology of V, and the embedding of V in its coordinate space, throughout a neighborhood of x^0. As an example, if K is a topological sphere, then V must be a topological manifold near x^0.

We will give the proof in some detail since similar methods will be important later, in Sections 4, 5, and 11.

Proof of 2.10. Again it is only necessary to consider the real case. Again let ε be small enough so that the disk D_ε contains no singular points of V, and no critical points of $r|(V - \Sigma(V))$, other than x^0 itself.

We will construct a smooth vector field $v(x)$ on the punctured disk $D_\varepsilon - x^0$ with two properties: The vector $v(x)$ will point "away" from x^0 for all x; that is the euclidean inner product

$$\langle v(x), x - x^0 \rangle$$

will be strictly positive. And the vector $v(x)$ will be tangent to the manifold $M_1 = V - \Sigma(V)$ whenever x belongs to M_1.

First let us construct such a vector field locally. Given any point x^α of $D_\varepsilon - x^0$ we will construct a vector field $v^\alpha(x)$ throughout a neighborhood U^α of x^α so that these two properties are satisfied.

If x^α does not belong to V then we can simply set

$$v^\alpha(x) = x - x^0$$

for all $\mathbf{x}$ in some neighborhood $U^\alpha \subset R^m - V$.

If $\mathbf{x}^\alpha$ belongs to V, and hence belongs to M_1, choose a system of local coordinates $u_1, \ldots, u_m$ throughout a neighborhood of $\mathbf{x}^\alpha$ so that M_1 corresponds to the locus $u_1 = \cdots = u_\rho = 0$. Since $\mathbf{x}^\alpha$ is not a critical point of the function $r \mid M_1$, where $r(\mathbf{x}) = \|\mathbf{x} - \mathbf{x}^0\|^2$, it follows that at least one of the partial derivatives

$$\partial r/\partial u_{\rho+1}, \ldots, \partial r/\partial u_m$$

must be non-zero at $\mathbf{x}^\alpha$. (Compare the proof of 2.7.) If for example $\partial r/\partial u_h$ is non-zero at $\mathbf{x}^\alpha$, then let U^α be a small connected neighborhood throughout which $\partial r/\partial u_h \neq 0$, and let $\mathbf{v}^\alpha(\mathbf{x})$ be the vector

$$\pm\ (\partial x_1/\partial u_h, \ldots, \partial x_m/\partial u_h)$$

tangent to the u_h-coordinate curve through $\mathbf{x}$; choosing the plus sign or the minus sign according as $\partial r/\partial u_h$ is positive or negative. This vector $\mathbf{v}^\alpha(\mathbf{x})$ is certainly tangent to M_1, whenever $\mathbf{x} \in M_1$, since the entire u_h-coordinate curve is contained in M_1. Furthermore

$$\begin{aligned} 2\langle \mathbf{v}^\alpha(\mathbf{x}), \mathbf{x} - \mathbf{x}^0 \rangle &= \sum 2(x_i - x_i^0) v_i^\alpha \\ &= \sum (\partial r/\partial x_i)(\pm\ \partial x_i/\partial u_h) \end{aligned}$$

is equal to $\pm\ \partial r/\partial u_h > 0$ for all $\mathbf{x} \in U^\alpha$.

Now choose a smooth partition of unity* $\{\lambda^\alpha\}$ on $D_\varepsilon - \mathbf{x}^0$, with Support $(\lambda^\alpha) \subset U^\alpha$. Then the vector field

$$\mathbf{v}(\mathbf{x}) = \sum \lambda^\alpha(\mathbf{x}) \mathbf{v}^\alpha(\mathbf{x})$$

---

* That is choose smooth real-valued functions λ^α on $D_\varepsilon - \mathbf{x}^0$ so that

$$\lambda^\alpha(\mathbf{x}) \geq 0, \qquad \Sigma_\alpha \lambda^\alpha(\mathbf{x}) = 1,$$

and so that each point of $D_\varepsilon - \mathbf{x}^0$ has a neighborhood within which only finitely many of the λ^α are non-zero. See for example DE RHAM, *Varietes differentiables* or LANG, *Differentiable manifolds*.

on $D_\varepsilon - \mathbf{x}^0$ clearly has the required properties.

Normalize by setting

$$\mathbf{w}(\mathbf{x}) = \mathbf{v}(\mathbf{x}) / \langle 2(\mathbf{x}-\mathbf{x}^0), \mathbf{v}(\mathbf{x}) \rangle$$

and consider the differential equation

$$d\mathbf{x}/dt = \mathbf{w}(\mathbf{x}) .$$

That is, look for smooth curves $\mathbf{x} = \mathbf{p}(t)$, defined say for $\alpha < t < \beta$, which satisfy

$$d\mathbf{p}(t)/dt = \mathbf{w}(\mathbf{p}(t)) .$$

Given any solution $\mathbf{p}(t)$, note that the derivative of the composition $r(\mathbf{p}(t))$ is given by

$$\begin{aligned} dr/dt &= \sum (\partial r/\partial x_i) w_i(\mathbf{x}) \\ &= \langle 2(\mathbf{x}-\mathbf{x}^0), \mathbf{w}(\mathbf{x}) \rangle = 1 \end{aligned}$$

where $\mathbf{x} = \mathbf{p}(t)$). So the function $r(\mathbf{p}(t))$ must be equal to t + constant. Thus, subtracting a constant from the parameter t if necessary, we may assume that

$$r(\mathbf{p}(t)) = \|\mathbf{p}(t) - \mathbf{x}^0\|^2 = t .$$

This solution $\mathbf{p}(t)$ can certainly be extended throughout the interval $0 < t \leq \varepsilon^2$.

[*Proof:* We may assume that the vector field $\mathbf{w}(\mathbf{x})$ has been constructed over an open set slightly larger than $D_\varepsilon - \mathbf{x}^0$, so that the boundary points of D_ε will not cause any trouble. By Zorn's lemma the given solution $\mathbf{p}(t)$ can be extended* over some maximal open interval $\alpha' < t < \beta'$. Suppose for example that $\beta' \leq \varepsilon^2$. Then we will extend the solution $\mathbf{p}(t)$ over a slightly larger interval, thus contradicting the definition of β'. Since the points $\mathbf{p}(t)$ with $\alpha' < t < \beta'$ all belong to the compact set D_ε, there exists at least one limit point $\mathbf{x}'$ of $\{\mathbf{p}(t)\}$ as $t \to \beta'$; and clearly $r(\mathbf{x}') = \beta' \neq 0$

* This use of Zorn's lemma could easily be eliminated.

so that $x' \in D_\varepsilon - x^0$. We will use the local existence, uniqueness and smoothness theorem* for the differential equation $dx/dt = w(x)$ near x'. This theorem asserts that for each x'' in some neighborhood U of x' and each t'' in some arbitrarily small open interval I containing β' there exists a unique solution

$$x = q(t), \qquad t \in I,$$

satisfying the initial condition $q(t'') = x''$; and furthermore, that $q(t)$ is a smooth function of x'', t'' and t. To apply this theorem, choose $t'' \in (\alpha', \beta') \cap I$ and let x'' equal $p(t'')$. Using the local uniqueness theorem we can verify that $p(t) = q(t)$ for all t in the common domain of definition $(\alpha', \beta') \cap I$. So the solutions p and q can be pieced together to yield a solution which is defined for all t in the larger interval $(\alpha', \beta') \cup I$. This contradiction proves that $\beta' > \varepsilon^2$; and a similar argument shows that $\alpha' = 0$.]

Note also that the solution $p(t)$, $0 < t \leq \varepsilon^2$ is uniquely determined by the initial value

$$p(\varepsilon^2) \in S_\varepsilon .$$

For each $a \in S_\varepsilon$ *let*

$$p(t) = P(a, t), \qquad 0 < t \leq \varepsilon^2,$$

be the unique solution which satisfies the initial condition

$$p(\varepsilon^2) = P(a, \varepsilon^2) = a .$$

Clearly this function P maps the product $S_\varepsilon \times (0, \varepsilon^2]$ diffeomorphically onto the punctured disk $D_\varepsilon - x^0$. Furthermore, since the vector field $w(x)$ is tangent to M_1 for all $x \in M_1$, it follows that every solution curve which touches M_1 must be contained in M_1. Hence P maps the product $K \times (0, \varepsilon^2]$ diffeomorphically onto $V \cap (D_\varepsilon - x^0)$.

Finally, note that $P(a, t)$ tends uniformly to x^0 as $t \to 0$. Therefore the correspondence

---

* See for example GRAVES, *Real Variables*, or LANG, *Differentiable Manifolds*.

$$ta + (1-t)x^0 \mapsto P(a, t\,\varepsilon^2) \;,$$

defined for $0 < t \leq 1$, extends uniquely to a homeomorphism from Cone(S_ε) to D_ε. Furthermore, this homeomorphism carries Cone(K) onto $V \cap D_\varepsilon$. This completes the proof of Theorem 2.10.

REMARK 2.11. It seems likely that Theorem 2.10 remains true even when x^0 is not an isolated singular point of V. Certainly it is known that every algebraic set can be triangulated, and hence that a suitably chosen neighborhood of any point is homeomorphic to the cone over something. See for example LOJASIEWICZ, *Triangulation of Semi-analytic Sets.*

The rest of §2 will study the rather dull case of a non-singular point of V, just to make sure that nothing unexpected happens.

LEMMA 2.12. *If* x^0 *is a simple point of* V, *then the intersection* $K = V \cap S_\varepsilon$ *is an unknotted sphere in* S_ε, *for all sufficiently small* ε.

Proof: Clearly the smooth function $r(x) = \|x - x^0\|^2$ restricted to $M_1 = V - \Sigma(V)$ has a non-degenerate critical point at x^0. Hence, by a lemma of MARSTON MORSE, there exists a system of local coordinates $u_1, \ldots, u_k$ for M_1 near x^0 so that

$$r(x) = u_1^2 + \cdots + u_k^2 \;.$$

(See for example MILNOR, *Morse Theory,* §2.2.) It follows immediately that $K = V \cap S_\varepsilon$ is diffeomorphic to the sphere consisting of all $(u_1, \ldots, u_k)$ with $u_1^2 + \cdots + u_k^2 = \varepsilon^2$.

But Morse's argument can be applied also the the pair of manifolds $M_1 \subset R^m$. That is: there exist local coordinates $u_1, \ldots, u_m$ for R^m near x^0 so that

$$r(x) = u_1^2 + \cdots + u_m^2 \;,$$

and so that V corresponds to the locus $u_{k+1} = \cdots = u_m = 0$. The proof of this sharpened form of Morse's lemma is a straightforward generalization, and will be left to the reader.

Thus the pair (S_ε, K) is diffeomorphic to the pair consisting of a sphere and a great sub-sphere in the u-coordinate space. This proves 2.12.

Now consider the special case of a simple point $\mathbf{z}^0$ of a complex hypersurface

$$V = f^{-1}(0) \subset C^{n+1} .$$

(Compare §1.) We want to study the set

$$F_0 = \phi^{-1}(1) = f^{-1}(R_+) \cap S_\varepsilon ,$$

where $\phi: S_\varepsilon - K \to S^1$ is defined by $\phi(\mathbf{z}) = f(\mathbf{z})/|f(\mathbf{z})|$.

LEMMA 2.13. *If the center* $\mathbf{z}^0$ *of* S_ε *is a regular point of* f, *then this "fiber"* F_0 *is diffeomorphic to* R^{2n}.

Proof: Applying Morse's argument to the pair of manifolds $V \subset f^{-1}(R)$ near $\mathbf{z}^0$ we obtain local coordinates $u_1, \ldots, u_{2n+1}$ for $f^{-1}(R)$ so that

$$\|\mathbf{z} - \mathbf{z}^0\|^2 = u_1^2 + \cdots + u_{2n+1}^2 ;$$

and so that V corresponds to the locus $u_1 = 0$. Then

$$\phi^{-1}(1) = f^{-1}(R_+) \cap S_\varepsilon$$

will correspond to the open hemisphere

$$\pm u_1 > 0, \; u_1^2 + u_2^2 + \cdots + u_{2n+1}^2 = \varepsilon^2 ;$$

which is clearly diffeomorphic to R^{2n}. This completes the proof.

Once we have proved the Fibration Theorem in §4, it will follow that the mapping

$$\phi: S_\varepsilon - K \to S^1$$

is the projection map of a trivial fiber bundle. In fact a theorem of T. E. Stewart implies that every smooth orientable bundle over S^1 with fiber diffeomorphic to euclidean space is trivial. See STEWART, Corollary 1; or note that any smooth bundle over S^1 is characterized by a certain diffeomorphism of the fiber (§8.4), and note that every diffeomorphism of euclidean space is smoothly isotopic either to the identity or to a reflection (STEWART, Theorem 1, or MILNOR, *Topology from the Differentiable Viewpoint*, p. 34.)

This completes the discussion in the case of a regular point.

§3. THE CURVE SELECTION LEMMA

The object of this section will be to prove the following.

Let $V \subset R^m$ be a real algebraic set, and let $U \subset R^m$ be an open set defined by finitely many polynomial inequalities:

$$U = \{x \in R^m \mid g_1(x) > 0, \ldots, g_\ell(x) > 0\} .$$

LEMMA 3.1. *If* $U \cap V$ *contains points arbitrarily close to the origin (that is if* $0 \in \text{Closure}(U \cap V)$*) then there exists a real analytic curve*

$$p : [0, \varepsilon) \to R^m$$

with $p(0) = 0$ and with $p(t) \in U \cap V$ *for* $t > 0$.

[Compare BRUHAT and CARTAN as well as WALLACE, *Algebraic Approximation*, §18.3.]

Proof: First suppose that the dimension of V is ≥ 2. Then we will construct a proper algebraic subset $V_1 \subset V$ so that $0 \in \text{Closure}(U \cap V_1)$. This procedure can then be iterated inductively until we find an algebraic subset V_q of dimension ≤ 1 with $0 \in \text{Closure}(U \cap V_q)$.

We may assume that V is irreducible. For if V is the union of two proper algebraic subsets, then one of these subsets will serve as V_1.

Also we may assume that the open set U does not contain any points of the singular set $\Sigma(V)$, within some neighborhood D_η of 0. For otherwise we could choose V_1 to be $\Sigma(V)$.

It will be convenient to use the language of differentials. By definition the *differential* $df(x)$ of a polynomial f at x is the element of the dual vector space

$$\mathrm{Hom}_R(R^m, R)$$

which corresponds to the row vector

$$(\partial f/\partial x_1, \ldots, \partial f/\partial x_m)$$

evaluated at $\mathbf{x}$.

Let $f_1, \ldots, f_k$ span the ideal $I(V)$. Recall that the singular set $\Sigma(V)$ is the set of all $\mathbf{x} \in V$ for which

$$\mathrm{rank}\{df_1(\mathbf{x}), \ldots, df_k(\mathbf{x})\} < \rho \;;$$

where the dimension of the variety V is $m - \rho$.

We will make use of two auxiliary functions:

$$r(\mathbf{x}) = \|\mathbf{x}\|^2, \qquad g(\mathbf{x}) = g_1(\mathbf{x})\, g_2(\mathbf{x}) \cdots g_\ell(\mathbf{x}) \;.$$

Let V' be the set of all $\mathbf{x} \in V$ with

$$\mathrm{rank}\{df_1(\mathbf{x}), \ldots, df_k(\mathbf{x}), dr(\mathbf{x}), dg(\mathbf{x})\} \le \rho + 1 \;.$$

We will prove:

LEMMA 3.2. *The intersection* $U \cap V'$ *also contains points arbitrarily close to* $\mathbf{0}$.

Proof: By hypothesis there exist arbitrarily small spheres S_ε centered at $\mathbf{0}$ which contain points of $U \cap V$. Choose any such sphere S_ε and consider the compact set consisting of all $\mathbf{x} \in V \cap S_\varepsilon$ with

$$g_1(\mathbf{x}) \ge 0, \ldots, g_\ell(\mathbf{x}) \ge 0 \;.$$

The continuous function g must be maximized at some point $\mathbf{x}'$ of this compact set; and clearly $\mathbf{x}' \in U$.

We will show that $\mathbf{x}' \in V'$.

Note first that S_ε intersects $U \cap V$ in a smooth manifold of dimension $m - \rho - 1$; and that

$$\mathrm{rank}\{df_1(\mathbf{x}), \ldots, df_k(\mathbf{x}), dr(\mathbf{x})\} = \rho + 1$$

at every point x of $U \cap V \cap S_\varepsilon$. This follows immediately from the proofs of 2.9, and 2.7, and the fact that $U \cap S_\varepsilon$ contains no singular points of V, providing that ε is small.

Now, proceeding as in §2.7, we see that the critical points of $g \mid U \cap V \cap S_\varepsilon$ are just those points of $U \cap V \cap S_\varepsilon$ which lie in V'. But $g \mid U \cap V \cap S_\varepsilon$ attains its maximum at x'. So x' is certainly a critical point; and therefore belongs to V'.

This completes the proof of 3.2.

Thus, if V' is a proper subset of V, then it satisfies our requirements. There remains the question of what to do in case $V = V'$.

We can also carry out the above construction using the function

$$(x_1, \ldots, x_m) \to x_i g(x_1, \ldots, x_m)$$

in place of g. Let V_i' be the set of all $x \in V$ with

$$\operatorname{rank}\{df_1(x), \ldots, df_k(x), dr(x), d(x_i g)(x)\} \leq \rho + 1 .$$

Then a similar argument shows that $0 \in \text{Closure}(U \cap V_i')$.

Thus we have found a suitable algebraic subset $V_1 \subset V$ except in the case

$$V = V' = V_1' = \cdots = V_m' .$$

ASSERTION: This case can only occur when the dimension $m - \rho$ of V is equal to 1.

Proof: We can certainly choose a point $x' \in U \cap V$ so that

$$\operatorname{rank}\{df_1(x'), \ldots, df_k(x'), dr(x')\} = \rho + 1 .$$

(Compare the proof of 3.2.) If $V = V'$, then $x' \in V'$ and hence the differential $dg(x')$ must belong to the $\rho + 1$ dimensional vector space spanned by

$$\{df_1(x'), \ldots, df_k(x'), dr(x')\} .$$

Similarly, if $V = V_i'$ then $d(x_i g)(x')$ must belong to this vector space. Using the identity

$$d(x_i g) = (dx_i)g + x_i(dg)$$

and the fact that $g(\mathbf{x}') \neq 0$ (since $\mathbf{x}' \in U$) it follows that $dx_i(\mathbf{x}')$ also belongs to this $\rho + 1$ dimensional vector space. But the differentials $dx_1, \ldots, dx_m$ form a basis for the entire m-dimensional vector space of differentials at $\mathbf{x}'$. So the subspace spanned by $df_1(\mathbf{x}'), \ldots, df_k(\mathbf{x}')$, and $dr(\mathbf{x}')$ must be the whole space; and its dimension $\rho + 1$ must be equal to m. This proves that the dimension $m - \rho$ of V is equal to 1.

Now we can make use of the classical local description of 1-dimensional varieties:

LEMMA 3.3. *Let $\mathbf{x}^0$ be a non-isolated point of a real (or complex) 1-dimensional variety* V. *Then a suitably chosen neighborhood of $\mathbf{x}^0$ in* V *is the union of finitely many "branches" which intersect only at $\mathbf{x}^0$. Each branch is homeomorphic to an open interval of real numbers (or to an open disk of complex numbers) under a homeomorphism $\mathbf{x} = \mathbf{p}(t)$ which is given by a power series*

$$\mathbf{p}(t) = \mathbf{x}^0 + \mathbf{a}_1 t + \mathbf{a}_2 t^2 + \mathbf{a}_3 t^3 + \cdots,$$

convergent for $|t| < \varepsilon$.

NOTE. Let k be the smallest index so that V is not contained in a coordinate hyperplane x_k = constant. Then the parametrization $\mathbf{p}$ can always be chosen so that $x_k = p_k(t)$ is a polynomial function of the form,

$$p_k(t) = \text{constant} \pm t^{\mu},$$

with* $\mu \geq 1$. Furthermore $\mathbf{p}$ can always be chosen so that the collection $\{i \mid \mathbf{a}_i \neq 0\}$ of exponents has greatest common divisor equal to 1. The power series $\mathbf{p}$ is then uniquely determined up to the sign of the parameter t (or up to multiplication of t by roots of unity in the complex case).

---

* Closely related is the Puiseux fractional power series expansion

$$\mathbf{x} = \mathbf{p}((\pm(x_k - x_k^0))^{1/\mu}).$$

Proof: For a complex curve in C^2 this is proved, for example, in VAN DER WAERDEN, *Algebraische Geometrie*, §14. The case of a complex curve in C^m, with $m > 2$, can be handled in exactly the same way.

The case of a real 1-dimensional variety $V \subset R^m$ can then be treated as follows. Let V_C be the smallest complex algebraic set in C^m which contains V. It is easily verified that V_C is irreducible, of complex dimension 1, and that the set $R^m \cap V_C$ of real points in V_C is equal to V. Now for each branch of V_C we can form the complex parametrization

$$\mathbf{x} = \mathbf{p}(t) = \mathbf{x}^0 + (0, \ldots, 0, t^\mu, \sum_i a_{k+1,i} t^i, \ldots, \sum_i a_{mi} t^i) .$$

We must then ask: for which values of the complex parameter t will the vector $\mathbf{p}(t)$ be real? Clearly the k-th component t^μ is real if and only if t can be expressed as the product of a 2μ-th root of unity ξ and a real number s. But, for each choice of ξ, substituting $t = \xi s$ in the power series $\mathbf{p}$, we obtain a new complex power series $\mathbf{x}^0 + \Sigma(\mathbf{a}_i \xi^i)s^i$ in the real variable s. If the coefficients $\mathbf{a}_i \xi^i$ are all real, then clearly $\mathbf{p}(\xi s) \in R^m$. But if some coefficient vector $\mathbf{a}_i \xi^i$ is not real, then $\mathbf{p}(\xi s) \notin R^m$ for all small non-zero values of s. Therefore each branch of V_C intersects R^m in at most a finite number of branches (actually in at most one branch) of the real variety V. This proves Lemma 3.3.

[REMARK. Presumably 3.3 remains true over any locally compact field; although the present proof does not apply.]

Now we are ready to finish the proof of the curve selection Lemma 3.1. Suppose that V contains points $\mathbf{x}$ arbitrarily close to $\mathbf{0}$ with $\mathbf{x} \in U$, that is with

$$g_1(\mathbf{x}) > 0, \ldots, g_\ell(\mathbf{x}) > 0 \;;$$

and suppose that V has dimension 1. Then one of the finitely many branches branches of V through $\mathbf{0}$ must contain points of U arbitrarily close to $\mathbf{0}$. Let

$$\mathbf{x} = \mathbf{p}(t), \qquad |t| < \varepsilon$$

be a real analytic parametrization of this branch. For each g_i note that the real analytic function $g_i(p(t))$ must be either > 0 for all t in some interval $0 < t < \varepsilon'$, or ≤ 0 for all t with $0 < t < \varepsilon'$. So the half-branch $p(0, \varepsilon')$ is either contained in U or disjoint from U, for ε' sufficiently small. Similarly the half-branch $p(-\varepsilon', 0)$ is contained in U or disjoint from U.

But we have assumed that $p(-\varepsilon, \varepsilon)$ contains points of U arbitrarily close to 0, so at least one of these two half-branches must be contained in U. This completes the proof of 3.1.

To conclude §3 let me give a typical application of Lemma 3.1 which will be useful in §11.

COROLLARY 3.4. *If* $f \geq 0$ *and* $g \geq 0$ *are non-negative polynomial functions on* R^m *which vanish at* x^0, *then for* x *belonging to some neighborhood* D_ε *of* x^0, *the two differentials* $df(x)$ *and* $dg(x)$ *cannot point in exactly opposite directions unless at least one of them vanishes.*

Proof: Let U be the open set consisting of all x for which the inner product

$$\sum_i (\partial f(x)/\partial x_i)(\partial g(x)/\partial x_i)$$

is negative, and let V be the algebraic set consisting of all x for which

$$\operatorname{rank}\{df(x), dg(x)\} \leq 1 .$$

Thus $U \cap V$ is the set of all x for which $df(x)$ and $dg(x)$ point in exactly opposite directions.

If $U \cap V$ contained points arbitrarily close to x^0, then there would exist an entire real analytic curve

$$x = p(t), \qquad 0 \leq t < \varepsilon ,$$

which consisted entirely of such points, except for $x^0 = p(0)$.

For every $x \in U$ note that $f(x) > 0$ and $g(x) > 0$. For if the non-negative function f were to vanish at x then the differential $df(x)$ would have to vanish also, hence **x** could not belong to U. Therefore

$$f(\mathbf{p}(t)) > 0 \text{ for } t > 0\ ;$$

and since $f \circ \mathbf{p}$ is real analytic this implies that $df(\mathbf{p}(t))/dt > 0$ also for small positive values of t. Similarly $dg(\mathbf{p}(t))/dt$ is positive for small positive t. But

$$df/dt = \sum (\partial f/\partial x_i) dp_i/dt, \qquad dg/dt = \sum (\partial g/\partial x_i) dp_i/dt\,,$$

where the row vector $(\partial f/\partial x_1, \ldots, \partial f/\partial x_m)$ is a negative multiple of $(\partial g/\partial x_1, \ldots, \partial g/\partial x_m)$, for all $t > 0$. Hence df/dt and dg/dt must have opposite sign.

This contradiction shows that the original premise must be false: $\mathbf{x}^0$ cannot be a limit point of $U \cap V$.

§4. THE FIBRATION THEOREM

Define the *gradient* of an analytic function $f(z_1, \ldots, z_m)$ of m complex variables to be the m-tuple

$$\mathbf{grad}\, f = (\overline{\partial f/\partial z_1}, \ldots, \overline{\partial f/\partial z_m})$$

whose j-th component is the complex conjugate of $\partial f/\partial z_j$. This definition is chosen so that the chain rule for the derivative of f along a path $\mathbf{z} = \mathbf{p}(t)$ takes the form

$$df(\mathbf{p}(t))/dt = \langle d\mathbf{p}/dt, \mathbf{grad}\, f\rangle ,$$

using the hermitian inner product

$$\langle \mathbf{a}, \mathbf{b}\rangle = \Sigma\, a_j \bar{b}_j$$

In other words the *directional derivative* of f along a vector $\mathbf{v}$ at the point $\mathbf{z}$ is equal to the inner product $\langle \mathbf{v}, \mathbf{grad}\, f(\mathbf{z})\rangle$.

Now let K denote the intersection of the set of zeros of f with the sphere S_ε, consisting of all $\mathbf{z}$ in C^m with $\|\mathbf{z}\| = \varepsilon$. Map the complement $S_\varepsilon - K$ to the unit circle S^1 by the correspondence

$$\phi(\mathbf{z}) = f(\mathbf{z})/|f(\mathbf{z})| .$$

LEMMA 4.1. *The critical points of this mapping* $\phi : S_\varepsilon - K \to S^1$ *are precisely those points* $\mathbf{z} \in S_\varepsilon - K$ *for which the vector* $i\, \mathbf{grad} \log f(\mathbf{z})$ *is a real multiple of the vector* $\mathbf{z}$.

(The logarithm of f is of course a many-valued function, but locally the logarithm can be defined as a single-valued function; and its gradient,

$$\mathbf{grad}\ \log f(\mathbf{z}) = (\mathbf{grad}\ f(\mathbf{z}))/\overline{f}(\mathbf{z}) ,$$

is well defined everywhere. Similar remarks apply to the function $\theta(\mathbf{z})$ considered next.)

Proof of 4.1. Setting $f(\mathbf{z})/|f(\mathbf{z})| = e^{i\theta(\mathbf{z})}$ note that the angle $\theta(\mathbf{z})$ can be described as the real part of $-i \log f(\mathbf{z})$. (To see this, multiply the equation

$$i\theta = \log(f/|f|) = \log f - \log |f|$$

by $-i$ and take the real part of both sides.) Differentiating the equation

$$\theta = \Re(-i \log f)$$

along a curve $\mathbf{z} = \mathbf{p}(t)$ we obtain

$$\begin{aligned} d\theta(\mathbf{p}(t))/dt &= \Re(d(-i \log f)/dt) \\ &= \Re\langle d\mathbf{p}/dt, \mathbf{grad}(-i \log f)\rangle \\ &= \Re\langle d\mathbf{p}/dt, i\, \mathbf{grad} \log f\rangle . \end{aligned}$$

In other words the directional derivative of the function $\theta(\mathbf{z})$ in the direction $\mathbf{v} = d\mathbf{p}/dt$ is equal to

$$\Re\langle \mathbf{v}, i\, \mathbf{grad} \log f(\mathbf{z})\rangle .$$

Now note that the hermitian vector space C^m can also be thought of as a euclidean vector space (of dimension 2m) over the real numbers, defining the euclidean inner product of two vectors $\mathbf{a}$ and $\mathbf{b}$ to be the real part

$$\Re\langle \mathbf{a}, \mathbf{b}\rangle = \Re\langle \mathbf{b}, \mathbf{a}\rangle .$$

Note for example that a vector $\mathbf{v}$ is tangent to the sphere S_ε at $\mathbf{z}$ if and only if the real inner product $\Re\langle \mathbf{v}, \mathbf{z}\rangle$ is zero.

Now if the vector $i\, \mathbf{grad}(\log f(\mathbf{z}))$ happens to be a real multiple of $\mathbf{z}$, (in other words if this vector is normal to S_ε) then for every vector $\mathbf{v}$ tangent to S_ε at $\mathbf{z}$ the directional derivative

$$\Re\langle \mathbf{v}, i\,\mathbf{grad}\log f(\mathbf{z})\rangle$$

of θ in the direction $\mathbf{v}$ will certainly be zero. Hence $\mathbf{z}$ is a critical point of the mapping ϕ.

On the other hand, if the vectors $i\,\mathbf{grad}\log f(\mathbf{z})$ and $\mathbf{z}$ are linearly independent over the real field, then there exists a vector $\mathbf{v}$ in our euclidean vector space so that

$$\Re\langle \mathbf{v}, \mathbf{z}\rangle = 0$$

$$\Re\langle \mathbf{v}, i\,\mathbf{grad}\log f(\mathbf{z})\rangle = 1 .$$

Thus $\mathbf{v}$ is tangent to S_ε and the directional derivative of θ along $\mathbf{v}$ is $+1 \neq 0$; hence $\mathbf{z}$ is not a critical point of ϕ. This proves 4.1.

Assume now that f *is a polynomial which vanishes at the origin.*

We want to prove that the associated mapping

$$\phi:\ S_\varepsilon - K \to S^1$$

has no critical points at all, for ε sufficiently small. In view of 4.1, we must prove the following. Let V denote the hypersurface $f^{-1}(0) \subset C^m$.

LEMMA 4.2. *For every* $\mathbf{z} \in C^m - V$ *which is sufficiently close to the origin, the two vectors* $\mathbf{z}$ *and* $i\,\mathbf{grad}\log f(\mathbf{z})$ *are linearly independent over* R.

In fact we will prove a slightly sharper statement:

LEMMA 4.3. *Given any polynomial* f *which vanishes at the origin, there exists a number* $\varepsilon_0 > 0$ *so that, for all* $\mathbf{z} \in C^m - V$ *with* $\|\mathbf{z}\| \leq \varepsilon_0$, *the two vectors* $\mathbf{z}$ *and* $\mathbf{grad}\log f(\mathbf{z})$ *are either linearly independent over the complex number or else*

$$\mathbf{grad}\log f(\mathbf{z}) = \lambda \mathbf{z}$$

where λ *is a non-zero complex number whose argument* has absolute value less than say* $\pi/4$.

* By the argument of $\lambda \neq 0$ will be meant the unique number $\theta \in (-\pi, \pi]$ such that $\lambda/|\lambda| = e^{i\theta}$.

In other words λ lies in the open quadrant of the complex plane which is centered about the positive real axis. *It follows that*

$$\Re(\lambda) > 0 ;$$

so that λ *cannot be pure imaginary.* Hence Lemma 4.2 will follow from 4.3.

The proof of 4.3 will depend on the curve selection lemma of §3 and on the following.

LEMMA 4.4. *Let* $p: [0, \varepsilon) \to C^m$ *be a real analytic path with* $p(0) = 0$ *such that, for each* $t > 0$, *the number* $f(p(t))$ *is non-zero and the vector* $\mathbf{grad} \log f(p(t))$ *is a complex multiple* $\lambda(t)p(t)$. *Then the argument of the complex number* $\lambda(t)$ *tends to zero as* $t \to 0$.

In other words $\lambda(t)$ is non-zero for small positive values of t and $\lim_{t\to 0} \lambda(t)/|\lambda(t)| = 1$.

Proof: Consider the Taylor expansions

$$\begin{aligned} p(t) &= at^{\alpha} + a_1 t^{\alpha+1} + a_2 t^{\alpha+2} + \cdots, \\ f(p(t)) &= bt^{\beta} + b_1 t^{\beta+1} + b_2 t^{\beta+2} + \cdots, \\ \mathbf{grad}\, f(p(t)) &= ct^{\gamma} + c_1 t^{\gamma+1} + c_2 t^{\gamma+2} + \cdots, \end{aligned}$$

where the leading coefficients a, b, c are non-zero. (The identity $df/dt = \langle dp/dt, \mathbf{grad}\, f\rangle$ shows that $\mathbf{grad}\, f(p(t))$ cannot be identically zero.) The leading exponents α, β, γ are integers with $\alpha \geq 1$, $\beta \geq 1$, and $\gamma \geq 0$. These series are all convergent say for $|t| < \varepsilon'$.

For each $t > 0$ we have

$$\mathbf{grad} \log f(p(t)) = \lambda(t)p(t) ,$$

hence

$$\mathbf{grad}\, f(p(t)) = \lambda(t)p(t)\overline{f}(p(t)) ,$$

or in other words

$$(ct^{\gamma} + \cdots) = \lambda(t)(a\bar{b}t^{\alpha+\beta} + \cdots) .$$

Comparing corresponding components of these two vector valued functions, we see that $\lambda(t)$ is a quotient of real analytic functions, and therefore has a Laurent expansion of the form

$$\lambda(t) = \lambda_0 t^{\gamma-\alpha-\beta}(1 + k_1 t + k_2 t^2 + \cdots) .$$

Furthermore the leading coefficients must satisfy the equation

$$\mathbf{c} = \lambda_0 \mathbf{a}\bar{\mathbf{b}} .$$

Substituting this equation in the power series expansion of the identity

$$df/dt = \langle d\mathbf{p}/dt, \mathbf{grad}\ f\rangle$$

we obtain

$$\begin{aligned}(\beta b t^{\beta-1} + \cdots) &= \langle \alpha \mathbf{a} t^{\alpha-1} + \cdots, \lambda_0 \mathbf{a}\bar{b} t^{\gamma} + \cdots \rangle \\ &= \alpha \|\mathbf{a}\|^2 \bar{\lambda}_0 b t^{\alpha-1+\gamma} + \cdots .\end{aligned}$$

Comparing the leading coefficients it follows that

$$\beta = \alpha \|\mathbf{a}\|^2 \bar{\lambda}_0$$

which proves that λ_0 is a positive real number. Therefore

$$\lim_{t \to 0} \text{ argument } \lambda(t) = 0 ,$$

which completes the proof of 4.4.

Proof of Lemma 4.3. First suppose that there were points $\mathbf{z} \in C^m - V$ arbitrarily close to the origin with

$$\mathbf{grad} \log f(\mathbf{z}) = \lambda \mathbf{z} \neq 0 ,$$

and with $|\text{argument } \lambda|$ strictly greater than $\pi/4$. In other words, assume that λ lies in the open half-plane

$$\Re((1 + i)\lambda) < 0$$

or the open half-plane

$$\Re((1-i)\lambda) < 0 .$$

We want to express these conditions by polynomial equalities and inequalities, so as to apply the curve selection lemma of §3.

Let W be the set of all $\mathbf{z}$ in C^m for which the vectors $\mathbf{grad}\, f(\mathbf{z})$ and $\mathbf{z}$ are linearly dependent. Thus $\mathbf{z} \in W$ if and only if the equations

$$z_j \overline{(\partial f/\partial z_k)} = z_k \overline{(\partial f/\partial z_j)}$$

are satisfied. Setting $z_j = x_j + iy_j$, and taking the real and imaginary parts, we obtain a collection of real polynomial equations in the real variables x_j and y_j. This proves that $W \subset C^m$ is a real algebraic set.

Note that a point $\mathbf{z} \in C^m - V$ belongs to W if and only if

$$(\mathbf{grad}\, f(\mathbf{z}))/\bar{f}(\mathbf{z}) = \lambda \mathbf{z}$$

for some complex number λ. Multiplying by $\bar{f}(\mathbf{z})$ and taking the inner product with $\bar{f}(\mathbf{z})\mathbf{z}$, this yields

$$<\mathbf{grad}\, f(\mathbf{z}), \bar{f}(\mathbf{z})\mathbf{z}> = \lambda \|\bar{f}(\mathbf{z})\mathbf{z}\|^2 .$$

In other words the number λ, multiplied by a positive real number, is equal to

$$\lambda'(\mathbf{z}) = <\mathbf{grad}\, f(\mathbf{z}), \bar{f}(\mathbf{z})\mathbf{z}> .$$

Hence

$$\text{argument } \lambda = \text{argument } \lambda' .$$

Clearly λ' is a (complex valued) polynomial function of the real variables x_j and y_j.

Now let U_+ (respectively U_-) denote the open set consisting of all $\mathbf{z}$ satisfying the real polynomial inequality

(*) $$\Re((1+i)\lambda'(\mathbf{z})) < 0$$

(respectively

$$\Re((1-i)\lambda'(\mathbf{z})) < 0 \text{).}$$

We have assumed that there exist points $\mathbf{z}$ arbitrarily close to the origin with $\mathbf{z} \in W \cap (U_+ \cup U_-)$. Hence, by 3.1, there must exist a real analytic path

$$\mathbf{p}: [0, \varepsilon) \to C^m$$

with $\mathbf{p}(0) = \mathbf{0}$ and with either

$$\mathbf{p}(t) \in W \cap U_+$$

for all $t > 0$, or

$$\mathbf{p}(t) \in W \cap U_-$$

for all $t > 0$. In either case, for each $t > 0$ we get

$$\mathbf{grad} \log f(\mathbf{p}(t)) = \lambda(t)\mathbf{p}(t)$$

with

$$|\text{argument } \lambda(t)| > \pi/4 \text{ ;}$$

which contradicts Lemma 4.4.

This contradiction does not quite complete the proof of 4.3. There remains the possibility that $W - (V \cap W)$ contains points $\mathbf{z}$ arbitrarily close to the origin with either $\lambda'(\mathbf{z}) = 0$ or

$$|\text{argument } \lambda'(\mathbf{z})| = \pi/4$$

But in this case we can go through essentially the same argument, substituting the polynomial equality

$$\Re((1+i)\lambda'(\mathbf{z}))\Re((1-i)\lambda'(\mathbf{z})) = 0 \text{ ,}$$

together with the polynomial inequality

$$\|f(\mathbf{z})\|^2 > 0 \text{ ,}$$

for the inequality (*). Again we would obtain a path $\mathbf{p}(t)$ which would contradict Lemma 4.4. This contradiction completes the proof of Lemmas 4.3 and 4.2.

Now, combining 4.1 and 4.2 we have proved:

COROLLARY 4.5. *If* $\varepsilon \leq \varepsilon_0$ *then the map*

$$\phi : S_\varepsilon - K \to S^1$$

has no critical points at all.

It follows that, for each $e^{i\theta} \in S^1$, the inverse image

$$F_\theta = \phi^{-1}(e^{i\theta}) \subset S_\varepsilon - K$$

is a smooth $(2m-2)$-dimensional manifold.

In order to prove that ϕ is actually the projection map of a locally trivial fibration we will need to make sharper use of 4.3 in order to carefully control the behavior of $\phi(\mathbf{z})$ as $\mathbf{z}$ tends to the set K where ϕ is not defined.

LEMMA 4.6. *If* $\varepsilon \leq \varepsilon_0$ *then there exists a smooth tangential vector field* $\mathbf{v}(\mathbf{z})$ *on* $S_\varepsilon - K$ *so that, for each* $\mathbf{z} \in S_\varepsilon - K$, *the complex inner product*

$$\langle \mathbf{v}(\mathbf{z}),\, i\, \mathbf{grad} \log f(\mathbf{z}) \rangle$$

is non-zero, and has argument less than $\pi/4$ *in absolute value.*

Proof: As in the proof of 2.10, it suffices to construct such a vector field locally, in the neighborhood of some given point $\mathbf{z}^\alpha$.

Case 1. If the vectors $\mathbf{z}^\alpha$ and $\mathbf{grad} \log f(\mathbf{z}^\alpha)$ are linearly independent over C, then the linear equations

$$\langle \mathbf{v}, \mathbf{z}^\alpha \rangle = 0 ,$$

$$\langle \mathbf{v},\, i\, \mathbf{grad} \log f(\mathbf{z}^\alpha) \rangle = 1$$

have a simultaneous solution $\mathbf{v}$. The first equation guarantees that

$\Re\langle \mathbf{v}, \mathbf{z}^a\rangle = 0$, so that $\mathbf{v}$ is tangent to S_ε at $\mathbf{z}^a$.

Case 2. If $\mathbf{grad} \log f(\mathbf{z}^a)$ is equal to a multiple $\lambda \mathbf{z}^a$, then set $\mathbf{v} = i\mathbf{z}^a$. Clearly

$$\Re\langle i\mathbf{z}^a, \mathbf{z}^a\rangle = 0 ;$$

and by 4.3 the number

$$\langle i\mathbf{z}^a, i\, \mathbf{grad} \log f(\mathbf{z}^a)\rangle = \bar{\lambda}\|\mathbf{z}^a\|^2$$

has argument less than $\pi/4$ in absolute value.

In either case one can choose a local tangential vector field $\mathbf{v}^a(\mathbf{z})$ which takes the constructed value $\mathbf{v}$ at $\mathbf{z}^a$. The condition

$$|\arg\langle \mathbf{v}^a(\mathbf{z}), i\, \mathbf{grad} \log f(\mathbf{z})\rangle| < \pi/4$$

will then certainly hold throughout a neighborhood of $\mathbf{z}^a$. Using a partition of unity, we obtain a global vector field $\mathbf{v}(\mathbf{z})$ having the same property. This proves 4.6.

Next normalize by setting

$$\mathbf{w}(\mathbf{z}) = \mathbf{v}(\mathbf{z})/\Re\langle \mathbf{v}(\mathbf{z}), i\, \mathbf{grad} \log f(\mathbf{z})\rangle .$$

Thus we obtain a smooth tangential vector field $\mathbf{w}$ on $S_\varepsilon - K$ which satisfies two conditions: *The real part of the inner product*

$$\langle \mathbf{w}(\mathbf{z}), i\, \mathbf{grad} \log f(\mathbf{z})\rangle$$

is identically equal to 1*; and the corresponding imaginary part satisfies*

$$|\Re\langle \mathbf{w}(\mathbf{z}), \mathbf{grad} \log f(\mathbf{z})\rangle| < 1 .$$

Consider the trajectories of the differential equation $d\mathbf{z}/dt = \mathbf{w}(\mathbf{z})$.

LEMMA 4.7. *Given any* $\mathbf{z}^0 \in S_\varepsilon - K$ *there exists a unique smooth path*

$$\mathbf{p} : R \to S_\varepsilon - K$$

which satisfies the differential equation

$$d\mathbf{p}/dt = \mathbf{w}(\mathbf{p}(t))$$

with initial condition $\mathbf{p}(0) = \mathbf{z}^0$.

Proof: Certainly such a solution $\mathbf{z} = \mathbf{p}(t)$ exists locally, and can be extended over some maximal open interval of real numbers. The only problem, which arises since $S_\varepsilon - K$ is non-compact, is to insure that $\mathbf{p}(t)$ cannot tend towards K as t tends toward some finite limit t_0. (Compare the proof of 2.10.) That is we must guarantee that $f(\mathbf{p}(t))$ cannot tend to zero, or

$$\mathfrak{R} \log f(\mathbf{p}(t)) \to -\infty ,$$

as $t \to t_0$. But the derivative

$$\begin{aligned} d\mathfrak{R} \log f/dt &= \mathfrak{R}\langle d\mathbf{p}/dt, \mathbf{grad} \log f\rangle \\ &= \mathfrak{R}\langle \mathbf{w}(\mathbf{p}(t)), \mathbf{grad} \log f\rangle \end{aligned}$$

has absolute value less than 1. Hence $|f(\mathbf{p}(t))|$ is bounded away from zero as t tends to any finite limit. This proves 4.7.

Setting $\phi(\mathbf{z}) = e^{i\theta(\mathbf{z})}$, as in the proof of 4.1, note that

$$d\theta(\mathbf{p}(t))/dt = \mathfrak{R}\langle d\mathbf{p}/dt, i\,\mathbf{grad} \log f\rangle$$

is identically equal to 1. Hence

$$\theta(\mathbf{p}(t)) = t + \text{constant} .$$

In other words the path $\mathbf{p}(t)$ projects under ϕ to a path which winds around the unit circle in the positive direction with unit velocity.

Clearly the point $\mathbf{p}(t)$ is a smooth function both of t and of the initial value

$$\mathbf{z}^0 = \mathbf{p}(0) .$$

Let us express this dependence by setting

$$p(t) = h_t(\mathbf{z}^0) .$$

Then each h_t is a diffeomorphism mapping $S_\varepsilon - K$ to itself, and mapping each fiber $F_\theta = \phi^{-1}(e^{i\theta})$ onto the fiber $F_{\theta+t}$. We now have no difficulty in proving the Fibration Theorem:

THEOREM 4.8. *For* $\varepsilon \leq \varepsilon_0$ *the space* $S_\varepsilon - K$ *is a smooth fiber bundle over* S^1, *with projection mapping* $\phi(\mathbf{z}) = f(\mathbf{z})/|f(\mathbf{z})|$.

Proof. Given $e^{i\theta} \in S^1$ let U be a small neighborhood of $e^{i\theta}$. Then the correspondence

$$(e^{i(t+\theta)}, \mathbf{z}) \mapsto h_t(\mathbf{z}) ,$$

for $|t| <$ constant, and $\mathbf{z} \in F_\theta$, maps the product $U \times F_\theta$ diffeomorphically onto $\phi^{-1}(U)$. This proves 4.8.

§5. THE TOPOLOGY OF THE FIBER AND OF K

We continue to study the locally trivial fibration

$$\phi: S_\varepsilon - K \to S^1$$

associated with a complex polynomial $f(z_1, \ldots, z_m)$ which vanishes at the origin. Setting $m = n + 1 \geq 1$ it follows from §4 that each fiber

$$F_\theta = \phi^{-1}(e^{i\theta})$$

is a smooth manifold of (real) dimension $2n$. This section will apply Morse theory to study the topology of F_θ and of K. The two principal results will be:

THEOREM 5.1. *Each fiber* F_θ *is parallelizable, and has the homotopy type of a finite* CW-*complex of dimension* n.

THEOREM 5.2. *The space* $K = V \cap S_\varepsilon$ *is* $(n-2)$*-connected.*

Thus for $n \geq 2$ the space K is connected, and for $n \geq 3$ it is simply connected. (For $n = 1$ a similar argument shows only that K is non-vacuous.)

The section concludes with an alternative description of the fibers: *Each* F_θ *is diffeomorphic to an open subset of a non-singular complex hypersurface, consisting of all* $\mathbf{z}$ *with* $\|\mathbf{z}\| < \varepsilon$ *and* $f(\mathbf{z}) =$ constant.

The proof of 5.1 will depend on a study of the Morse theory associated with the smooth real-valued function $|f|$ on F_θ. The proof of 5.2 will depend on a parallel study of the smooth function $|f|$ on the entire total space $S_\varepsilon - K$. In each case we will show that the Morse index* of $|f|$ at any critical point is $\geq n$.

As a first step it is necessary to identify the critical points. It will be convenient to work rather with the smooth functions $a_\theta : F_\theta \to R$ and $a : S_\varepsilon - K \to R$, defined by

$$a_\theta(\mathbf{z}) = a(\mathbf{z}) = \log|f(\mathbf{z})| .$$

Clearly the critical points of a are the same as those of $|f|$ on $S_\varepsilon - K$, and similarly the critical points of a_θ are the same as those of $|f|$ restricted to F_θ.

LEMMA 5.3. *The critical points of the smooth real-valued function* $a_\theta(\mathbf{z}) = \log|f(\mathbf{z})|$ *on* F_θ *are those points* $\mathbf{z} \in F_\theta$ *for which the vector* **grad** $\log f(\mathbf{z})$ *is a complex multiple of* $\mathbf{z}$.

Proof: Proceeding as in the proof of 4.1 note that the directional derivative of the function

$$\log|f(\mathbf{z})| = \Re \log f(\mathbf{z})$$

in any direction $\mathbf{v}$ is equal to the real inner product

$$\Re\langle \mathbf{v}, \mathbf{grad} \log f(\mathbf{z})\rangle .$$

Thus $\mathbf{z}$ will be a critical point of this function, restricted to F_θ, if and only if the vector **grad** $\log f(\mathbf{z})$ is normal to F_θ at $\mathbf{z}$. (Here "normal" means "orthogonal to all tangent vectors," using the real inner product.)

But, comparing the proof of 4.1, we see that the space of normal vectors to the submanifold $F_\theta \subset C^m$, of real codimension 2, is spanned by the two independent vectors $\mathbf{z}$ and $i\,\mathbf{grad} \log f(\mathbf{z})$. Thus $\mathbf{z}$ is a critical point of a_θ if and only if there is a real linear relation between the vectors **grad** $\log f(\mathbf{z})$, $\mathbf{z}$ and $i\,\mathbf{grad} \log f(\mathbf{z})$. Clearly this proves 5.3.

---

* See for example MILNOR, *Morse Theory*, §2.

REMARK 5.4. Note that the tangent space of F_θ at a critical point $\mathbf{z}$ of a_θ is actually a complex vector space, consisting of all $\mathbf{v}$ with $\langle \mathbf{v}, \mathbf{z} \rangle = 0$. For if $i\,\mathbf{grad} \log f(\mathbf{z})$ is a complex multiple of $\mathbf{z}$, then a vector $\mathbf{v}$ is real orthogonal to both $\mathbf{z}$ and $i\,\mathbf{grad} \log f(\mathbf{z})$ if and only if it is complex orthogonal to $\mathbf{z}$.

Next we must study the Hessian of the smooth function a_θ at a critical point, in order to compute the Morse index. We will use the following interpretation of the Hessian. Given a tangent vector $\mathbf{v}$ at the critical point $\mathbf{z}$ choose a smooth path

$$\mathbf{p}: R \to F_\theta$$

with velocity vector $d\mathbf{p}/dt = \mathbf{v}$ at $\mathbf{p}(0) = \mathbf{z}$. Then the second derivative

$$\ddot{a}_\theta = d^2 a_\theta(\mathbf{p}(t))/dt^2$$

at $t = 0$ can be expressed as a quadratic function of $\mathbf{v}$, and this quadratic function is the Hessian.

LEMMA 5.5 *The second derivative of* $a_\theta(\mathbf{p}(t))$ *at* $t = 0$ *is given by an expression of the form*

$$\ddot{a}_\theta = \sum \mathcal{R}(b_{jk} v_j v_k) - c\,\|\mathbf{v}\|^2 ,$$

where (b_{jk}) *is a matrix of complex numbers and* c *is a positive real number.*

Proof: Note first that our path $\mathbf{p}(t)$ lies within the manifold F_θ on which $f/|f| = e^{i\theta}$ is constant. Differentiating the identity

$$a_\theta(\mathbf{p}(t)) = \log |f(\mathbf{p}(t))| = \log f(\mathbf{p}(t)) - i\theta$$

we obtain

$$\dot{a}_\theta = d \log f/dt = \sum (\partial \log f/\partial z_j)(dp_j/dt) .$$

Differentiating again:

$$\ddot{a}_\theta = \sum (\partial \log f/\partial z_j)(d^2 p_j/dt^2)$$
$$+ \sum (\partial^2 \log f/\partial z_j \partial z_k)(dp_j/dt)(dp_k/dt) .$$

Setting $t = 0$, setting

$$\mathbf{grad} \log f(\mathbf{z}) = \lambda \mathbf{z}$$

(by 5.3), and introducing the abbreviation

$$D_{jk} = \partial^2 \log f/\partial z_j \partial z_k ,$$

this can be written as

$$\ddot{a}_\theta = \langle \ddot{\mathbf{p}}, \lambda \mathbf{z} \rangle + \sum D_{jk} v_j v_k ,$$

where the left side $\ddot{a}_\theta$ is clearly real. Now multiply both sides by λ and take the real part:

$$\ddot{a}_\theta \Re(\lambda) = |\lambda|^2 \Re \langle \ddot{\mathbf{p}}, \mathbf{z} \rangle + \sum \Re(\lambda D_{jk} v_j v_k) .$$

Substituting the identity

$$\Re \langle \ddot{\mathbf{p}}, \mathbf{z} \rangle = -\|\mathbf{v}\|^2 ,$$

which is obtained by twice differentiating the equation

$$\langle \mathbf{p}(t), \mathbf{p}(t) \rangle = \text{constant} ,$$

into this, we obtain

$$\ddot{a}_\theta \Re(\lambda) = \sum \Re(\lambda D_{jk} v_j v_k) - \|\lambda \mathbf{v}\|^2 .$$

Dividing by $\Re(\lambda)$, which is positive by 4.3, this completes the proof of Lemma 5.5.

Now we can easily estimate the index.

LEMMA 5.6. *The Morse index of* $a_\theta \colon F_\theta \to R$ *at a critical point is* $\geq n$. *Hence the Morse index of* $a\colon S_\varepsilon - K \to R$ *at any critical point is also* $\geq n$.

Proof: The Morse Index I of the quadratic function

$$H(\mathbf{v}) = \Re\left(\sum b_{jk} v_j v_k\right) - c\,\|\mathbf{v}\|^2 ,$$

where $\mathbf{v}$ ranges over the tangent space of F_θ at $\mathbf{z}$, is defined as the maximum dimension of a linear subspace on which H is negative definite.

If $H(\mathbf{v}) \geq 0$ for any non-zero vector $\mathbf{v}$, note that $H(i\mathbf{v}) < 0$; for the first term in our expression for $H(\mathbf{v})$ changes sign and the second term remains negative. Note that $i\mathbf{v}$ is also a tangent vector to F_θ, by Remark 5.4.

Now split the tangent space at $\mathbf{z}$ as a real direct sum $T_0 \oplus T_1$ where the Hessian H is negative definite on T_0 and positive semi-definite on T_1. Clearly the dimension of T_0 is equal to the Morse index I.

But H is also negative definite on iT_1. Therefore

$$I \geq \dim(iT_1) = \dim T_1 = 2n - I .$$

In other words $I \geq n$, which proves the first part of 5.6.

The corresponding statement for $a\colon S_\varepsilon - K \to R$ follows immediately. Every critical point of a is also a critical point of the appropriate a_θ, and the index of a at $\mathbf{z}$ is clearly greater than or equal to the index of a_θ at $\mathbf{z}$. This proves 5.6.

Next we must verify that the critical points all lie within a compact subset of F_θ (or of $S_\varepsilon - K$).

LEMMA 5.7. *There exists a constant* $\eta_\theta > 0$ *so that the critical points of* a_θ *all lie within the compact subset* $|f(\mathbf{z})| \geq \eta_\theta$ *of* F_θ. *Similarly, there exists* $\eta > 0$ *so that the critical points* $\mathbf{z}$ *of* a *all satisfy* $|f(\mathbf{z})| \geq \eta$.

This is proved using either 2.8 or 3.1. For example if there were critical points $\mathbf{z}$ of $a_\theta = \log|f|$ on F_θ with $|f(\mathbf{z})|$ arbitrarily close to zero, then these critical points would have a limit point $\mathbf{z}^0$ on the compact set S_ε. Using the Curve Selection Lemma, there would exist a smooth path

$$p : (0, \varepsilon') \to F_\theta$$

consisting completely of critical points, with

$$p(t) \to \mathbf{z}^0 \text{ as } t \to 0 .$$

Clearly the function a_θ is constant along this path, hence $|f|$ is constant, and cannot tend to $|f(\mathbf{z}^0)| = 0$. This contradiction proves 5.7.

LEMMA 5.8. *There exists a smooth mapping*

$$s_\theta : F_\theta \to R_+$$

so that all critical points of s_θ *are non-degenerate, with Morse index* $\geq n$, *and so that*

$$s_\theta(\mathbf{z}) = |f(\mathbf{z})|$$

whenever $|f(\mathbf{z})|$ *is sufficiently close to zero. Similarly there exists a smooth mapping* s *from* $S_\varepsilon - K$ *to the positive reals with all critical points non-degenerate, of index* $\geq n$, *and with*

$$s(\mathbf{z}) = |f(\mathbf{z})|$$

whenever $|f(\mathbf{z})|$ *is sufficiently close to zero.*

Proof: According to MORSE, Theorem 8.7, p. 178, we can choose s_θ (or s) so as to coincide with $|f|$ except on a compact neighborhood of the critical set, so as to have only non-degenerate critical points, and so that the first and second derivatives of s_θ on any compact coordinate patch uniformly approximate those of $|f|$. Since the critical points of $|f|$ all have index $\geq n$, it follows, if the approximation is sufficiently close, that the critical points of s_θ also have index $\geq n$. (See for example MILNOR, *Morse Theory*, §22.4.) This completes the proof of 5.8.

Note that the critical points of s_θ are isolated, and all lie within a compact set. Hence there are only finitely many critical points.

We are now ready to prove Theorems 5.1 and 5.2.

Proof of 5.1. In order to apply Morse theory in its usual form we need a non-degenerate mapping $g: F_\theta \to R$ with the property that the set of $\mathbf{z}$ with $g(\mathbf{z}) \leq c$ is compact, for every constant c. (In other words g should be proper and bounded from below.) Clearly the function

$$g(\mathbf{z}) = -\log s_\theta(\mathbf{z})$$

satisfies these conditions.

The index I of s_θ or of $\log s_\theta$ at a critical point is $\geq n$. Hence the index of $-\log s_\theta$ is $2n - I \leq n$. Now according to the main theorem of Morse theory (see MILNOR, *Morse Theory*, §3.5) the manifold F_θ has the homotopy type of a CW-complex of dimension $\leq n$, made up of one cell for each critical point of g. This proves half of 5.1.

In particular, if $n \geq 1$, it follows that the homology group $H_{2n}(F_\theta; Z_2)$ is zero. So the 2n-dimensional manifold F_θ cannot have any compact component.

To complete the proof of 5.1 it is only necessary to verify that F_θ is parallelizable. But F_θ is embedded in the sphere S_ε, and hence in the coordinate space C^{n+1}, with trivial normal bundle. Since F_θ has no compact component, it follows that F_θ is parallelizable. (See KERVAIRE and MILNOR, §3.4.) This completes the proof.

REMARK. A similar argument shows that the total space $S_\varepsilon - K$ has the homotopy type of a finite complex of dimension $n + 1$.

Proof of 5.2. Let $N_\eta(K)$ denote the neighborhood of K consisting of all $\mathbf{z} \in S_\varepsilon$ with $|f(\mathbf{z})| \leq \eta$. It follows from 5.7 that $N_\eta(K)$ is a smooth manifold with boundary, for η sufficiently small. Using the smooth non-degenerate real.valued function s on S_ε – Interior $N_\eta(K)$ we see that the entire sphere S_ε has the homotopy type of a complex built up from $N_\eta(K)$ by adjoining finitely many cells of dimension $\geq n$, one I-cell for each critical point of s of index I. (Compare *Morse Theory*, §3. In fact, according to SMALE one has the more precise description that the smooth manifold S_ε can be built up from $N_\eta(K)$ by adjoining finitely many "handles," each handle having index $\geq n$.)

Clearly the adjunction of a cell of dimension $\geq n$ cannot alter the homotopy groups in dimension $\leq n-2$. Therefore

$$\pi_i(N_\eta(K)) \cong \pi_i(S_\varepsilon) = 0$$

for $i \leq n-2$.

To complete the proof we must make use of the fact that K is an absolute neighborhood retract. In fact K is a real algebraic set and hence, according to LOJASIEWICZ, can actually be triangulated.

Therefore K is a retract of the neighborhood $N_\eta(K)$ providing that η is sufficiently small.* It follows that $\pi_i(K)$ is also trivial for $i \leq n-2$, which completes the proof of 5.2.

To conclude §5 we give an alternative description of the fibers. First two lemmas. Let D_ε denote the closed disk bounded by S_ε.

LEMMA 5.9. *There exists a smooth vector field* $\mathbf{v}$ *on* $D_\varepsilon - V$ *so that the inner product*

$$\langle \mathbf{v}(\mathbf{z}), \operatorname{grad} \log f(\mathbf{z}) \rangle$$

is real and positive, for all $\mathbf{z}$ *in* $D_\varepsilon - V$, *and so that the inner product* $\langle \mathbf{v}(\mathbf{z}), \mathbf{z} \rangle$ *has positive real part.*

The proof is similar to that of 4.6, and will be left to the reader.

* See for example HU.

Now consider the solutions of the differential equation

$$dp/dt = v(p(t))$$

on $D_\varepsilon - V$. The condition that

$$\langle dp/dt, \mathbf{grad} \log f(p(t))\rangle$$

is real and positive tells us that the argument of $f(p(t))$ is constant, and that $|f(p(t))|$ is strictly monotone as a function of t. The condition

$$2\Re\langle dp/dt, p(t)\rangle = d\|p(t)\|^2/dt > 0$$

guarantees that $\|p(t)\|$ is a strictly monotone function of t.

Thus starting at any interior point $\mathbf{z}$ of $D_\varepsilon - V$ and following the trajectory through $\mathbf{z}$ we travel "away" from the origin, in a direction of increasing $|f|$, until we reach a point $\mathbf{z}'$ on $S_\varepsilon - K$, which must satisfy

$$f(\mathbf{z}')/|f(\mathbf{z}')| = f(\mathbf{z})/|f(\mathbf{z})| .$$

Using this correspondence $\mathbf{z} \mapsto \mathbf{z}'$ we clearly prove the following. Let c be a small complex constant, and let $c/|c| = e^{i\theta}$. (Compare Figure 4.)

LEMMA 5.10. *The intersection of the hyperplane* $f^{-1}(c)$ *with the open* ε*-disk is diffeomorphic to the portion of the fiber* F_θ *defined by the inequality* $|f(\mathbf{z})| > |c|$.

But if $|c|$ is sufficiently small, then it follows from 5.7 that this portion of F_θ is diffeomorphic to all of F_θ. (Compare *Morse Theory*, §3.1.) Thus we have proved:

THEOREM 5.11. *If the complex number* $c \neq 0$ *is sufficiently close to zero, then the complex hypersurface* $f^{-1}(c)$ *intersects the open* ε*-disk in a smooth manifold which is diffeomorphic to the fiber* F_θ.

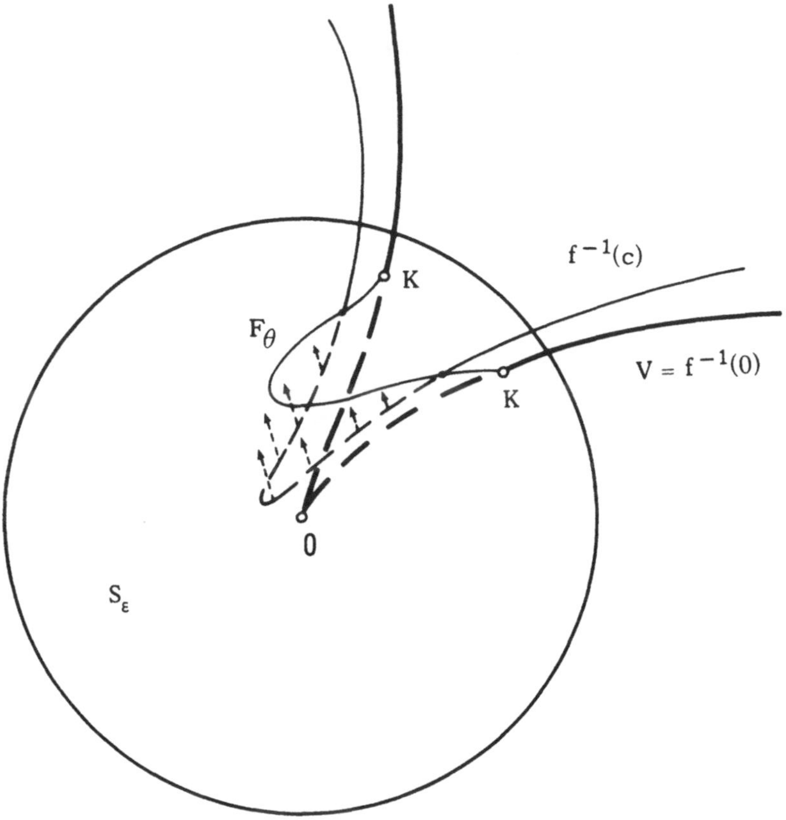

Figure 4.

REMARK. If we combine 5.11 with ANDREOTTI and FRANKEL'S analysis of the Morse theory of the smooth real-valued function $\|\mathbf{z}\|^2$ on $f^{-1}(c)$, then it is possible to obtain an alternate proof of Theorem 5.1.

§6. THE CASE OF AN ISOLATED CRITICAL POINT

Now make the additional hypothesis that the polynomial $f(z_1, \ldots, z_{n+1})$ has no critical points in some neighborhood of the origin, except possibly for the origin itself. Thus the origin is either an isolated singular point, or a non-singular point, of the hypersurface $V = f^{-1}(0)$. (Compare §2.5.) Assume also that $n \geq 1$.

According to §2.9 the intersection $K = V \cap S_\varepsilon$ is a smooth $(2n-1)$-dimensional manifold, providing that ε is sufficiently small. This statement can be sharpened as follows.

LEMMA 6.1. *For ε sufficiently small, the closure of each fiber F_θ in S_ε is a smooth $2n$-dimensional manifold with boundary, the interior of this manifold being F_θ and the boundary being precisely K.*

Proof: First note that the mapping $f | S_\varepsilon$ to C has no critical points on K, for ε sufficiently small. (In other words, the number zero is a regular value of $f | S_\varepsilon$.) This can be derived from the proof of 2.9, or can be proved as follows, using the Curve Selection Lemma. The critical points of $f | S_\varepsilon$ are clearly those points $\mathbf{z}$ in S_ε at which the (non-zero) vector $\mathbf{grad}\, f(\mathbf{z})$ is a complex multiple of $\mathbf{z}$. Given a non-smooth path

$$\mathbf{p}: [0, \varepsilon') \to C^{n+1}$$

consisting exclusively of such points, with

$$\mathbf{p}(0) = \mathbf{0} \quad \text{and} \quad f(\mathbf{p}(t)) \equiv 0 ,$$

we would have

$$< d\mathbf{p}/dt, \mathbf{grad}\, f > \; = \; df(\mathbf{p}(t))/dt \equiv 0$$

hence

$$2\Re\langle dp/dt, p(t)\rangle = d\|p(t)\|^2/dt \equiv 0$$

and therefore $p(t) \equiv 0$, which contradicts the hypothesis.

Now let $\mathbf{z}^0$ be any point of K. Choose (real) local coordinates $u_1, \ldots, u_{2n+1}$ for S_ε in a neighborhood U of $\mathbf{z}^0$ so that

$$f(\mathbf{z}) = u_1(\mathbf{z}) + iu_2(\mathbf{z})$$

for all $\mathbf{z} \in U$. Note that a point of U belongs to the fiber $F_0 = \phi^{-1}(1)$ if and only if

$$u_1 > 0, \quad u_2 = 0.$$

Hence the closure $\overline{F}_0$ intersects U in the set

$$u_1 \geq 0, \quad u_2 = 0.$$

Clearly this is a smooth 2n-dimensional manifold, with $F_0 \cap U$ as interior and with $K \cap U$ as boundary.

This discussion for other fibers F_θ is similar. This completes the proof.

COROLLARY 6.2. *The compact manifold-with-boundary* $\overline{F}_\theta$ *is embedded in* S_ε *in such a way as to have the same homotopy type as its complement* $S_\varepsilon - \overline{F}_\theta$.

For the complement is a locally trivial fiber space over the contractible manifold $S^1 - (e^{i\theta})$. Hence $S_\varepsilon - \overline{F}_\theta$ has any other fiber $F_{\theta'}$, as deformation retract. But $F_{\theta'}$ is diffeomorphic to F_θ, and so has the same homotopy type as $\overline{F}_\theta$.

COROLLARY 6.3. *The fiber* F_θ *has the homology of a point in dimensions less than* n.

(Compare FARY, p. 31.)

This follows from the Alexander duality theorem which says that the reduced* homology group $\tilde{H}_i(S_\varepsilon - \overline{F}_\theta)$ is isomorphic to the reduced cohomology group $\tilde{H}^{2n-i}(\overline{F}_\theta)$, which is zero for $2n - i > n$ by 5.1.

The statement can be sharpened as follows:

LEMMA 6.4. *The fiber* F_θ *is actually* $(n-1)$*-connected.*

Proof: In view of 6.3 we need only verify that F_θ is simply connected, provided that $n \geq 2$.

For $n \geq 3$ this can be proved using Lemma 5.8. Using the smooth function s_θ on $\overline{F}_\theta$ note that $\overline{F}_\theta$ can be constructed, starting with a neighborhood $K \times [0, \eta]$ of its boundary by adjoining handles of index $\geq n$, there being one handle corresponding to each critical point of s_θ. (Compare the proof of 5.2.) Since the adjunction of such handles cannot change the homotopy groups in dimensions $\leq n-2$, it follows that

$$\pi_i(\overline{F}_\theta) \cong \pi_i(K \times [0, \eta]) = 0$$

for $i \leq n-2$, making use of Theorem 5.2.

Here is a better argument, which works also in the case $n = 2$. Using the Morse function $-s_\theta$ on $\overline{F}_\theta$, note that $\overline{F}_\theta$ can be built up by starting with a disk D_0^{2n} and successively adjoining handles of index $\leq n$. All of these handles can be attached within the containing space S_ε. But the complement $S_\varepsilon - D_0^{2n}$ is certainly simply connected, and the adjunction of handles of index $\leq \dim(S_\varepsilon) - 3 = 2n-2$ cannot change the fundamental group of the complementary set. So it follows inductively that the complement $S_\varepsilon - \overline{F}_\theta$ is also simply connected, provided that $n \leq 2n-2$. Together with 6.2, this completes the proof.

THEOREM 6.5. *Each fiber has the homotopy type of a bouquet* $S^n \vee S^n \vee \cdots \vee S^n$ *of spheres.*

---

* The reduced group $\tilde{H}_i X$ is defined to be the kernel of the natural homomorphism $H_i X \to H_i(\text{point})$. Similarly $\tilde{H}^i X$ is the cokernel of $H^i(\text{point}) \to H^i X$.

Proof: The homology group* $H_n(F_\theta)$ must be free abelian, since any torsion elements would give rise to cohomology classes in dimension $n+1$, contradicting 5.1. Hence $\pi_n(F_\theta) \cong H_n(F_\theta)$ is free abelian, using the Hurewicz theorem and assuming $n \geq 2$; so we can choose finitely many maps

$$(S^n, \text{base point}) \longrightarrow (F_\theta, \text{base point})$$

representing a basis. These combine to yield a map

$$S^n \vee \cdots \vee S^n \to F_\theta$$

which induces an isomorphism of homology groups and hence, by Whitehead's theorem, is a homotopy equivalence. This completes the proof for the case $n \geq 2$, The proof for $n = 1$ will be left to the reader.

We will see in §7 that the number of spheres in this bouquet is never zero, unless the origin is a regular point of f.

For $n \neq 2$ a still sharper statement is possible

THEOREM 6.6. *For* $n \neq 2$ *the manifold* $\bar{F}_\theta$ *is diffeomorphic to a handlebody, obtained from the disk* D^{2n} *by simultaneously attaching a number of handles of index precisely equal to* n.

Proof: This follows from SMALE, "On the structure of 5-manifolds," §1.2, together with our 5.2 and 6.4.

For a thorough discussion of such handlebodies, the reader is referred to WALL.

It may be conjectured that Theorem 6.6 remains true when $n = 2$.

---

* All homology groups are to have integer coefficients, unless otherwise specified.

§7. THE MIDDLE BETTI NUMBER OF THE FIBER

An isolated critical point $\mathbf{z}^0$ of a complex polynomial $f(z_1, \ldots, z_m)$ is called *non-degenerate* if the Hessian matrix $(\partial^2 f/\partial z_j \partial z_k)$ at $\mathbf{z}^0$ is non-singular. Otherwise $\mathbf{z}^0$ is a *degenerate* critical point.

We will introduce a positive integer μ which measures the amount of degeneracy at the critical point $\mathbf{z}^0$. This integer μ will be described as the multiplicity of $\mathbf{z}^0$ as solution to the collection of polynomial equations

$$\partial f/\partial z_1 = \cdots = \partial f/\partial z_m = 0 .$$

Consider first a somewhat more general situation. Let

$$g_1(\mathbf{z}), \ldots, g_m(\mathbf{z})$$

be arbitrary analytic functions of m complex variables, and let $\mathbf{z}^0$ be an isolated solution to the collection of equations

$$g_1(\mathbf{z}) = \cdots = g_m(\mathbf{z}) = 0 .$$

Briefly we will say that $\mathbf{z}^0$ is an isolated "zero" of the mapping $\mathbf{g} : \mathbf{C}^m \to \mathbf{C}^m$.

Definition. The multiplicity μ of the isolated zero $\mathbf{z}^0$ is the degree of the mapping

$$\mathbf{z} \mapsto \mathbf{g}(\mathbf{z})/\|\mathbf{g}(\mathbf{z})\|$$

from a small sphere S_ε centered at $\mathbf{z}^0$ to the unit sphere of $\mathbf{C}^m$.

Presumably this definition agrees with the various definitions used by algebraic geometers. (See for example VAN DER WAERDEN, *Algebraische Geometrie*, §38, or HODGE and PEDOE, p. 120-129.) But the topological definition is more convenient for our purpose.

The following result helps to justify this definition.

THEOREM 7.1 (Lefschetz). *The multiplicity* μ *is always a positive integer.*

The proof of 7.1, and further discussion of μ, will be given in Appendix B. (See also LEFSCHETZ, *Topology*, p. 382.)

Now let us return to the situation of §6. Let the origin be an isolated critical point, and a zero, of the polynomial $f(z_1, \ldots, z_{n+1})$. Thus the equations

$$\partial f/\partial z_1 = \cdots = \partial f/\partial z_{n+1} = 0$$

have an isolated solution $\mathbf{z} = 0$. Let μ be the multiplicity of this solution. As in the preceding sections, we consider the associated fibration having the 2n-manifold

$$F_0 = \{\mathbf{z} \in S_\varepsilon \mid f(\mathbf{z}) > 0\}$$

as a typical fiber. The main result of this section will be:

THEOREM 7.2. *The middle Betti number of the fiber* F_0 *is equal to the multiplicity* μ. *Hence the middle homology group* $H_n F_0$ *is free abelian of rank* μ.

(REMARK. Brieskorn has recently given a simple proof of Theorem 7.2 which is quite different from the one presented here.)

Since $\mu > 0$ by 7.1, this implies:

COROLLARY 7.3. *If the origin is an isolated critical point of* f, *then the fibers* F_θ *are not contractible, and the manifold* $K = V \cap S_\varepsilon$ *is not an unknotted sphere in* S_ε.

(This contrasts with §2.13 and §2.12 which showed that F_θ is contractible, and that $K \subset S_\varepsilon$ is an unknotted sphere, whenever the origin is a regular point of f.)

For if K were a topologically unknotted sphere in S_ε, then $S_\varepsilon - K$ would have the homotopy type of a circle. The homotopy exact sequence

$$\cdots \to \pi_{n+1}(S^1) \to \pi_n(F_0) \to \pi_n(S_\varepsilon - K) \to \cdots$$

of the fibration would then lead to a contradiction (even when $n = 1$).

In order to prove 7.2 we will need a device for computing the degree of a smooth map

$$\mathbf{v}: \; S^k \to S^k$$

of a sphere into itself in terms of the fixed points of $\mathbf{v}$. Let M be a compact region with smooth boundary on the sphere $S^k \subset R^{k+1}$, and for each boundary point $\mathbf{x}$ of M let $\mathbf{n}(\mathbf{x})$ denote the *inward normal vector,* the unique unit vector which is tangent to S^k and normal to ∂M at $\mathbf{x}$, and points into M.

LEMMA 7.4. *If* (1) *every fixed point of the mapping* $\mathbf{v}: S^k \to S^k$ *lies in the interior of* M, *if* (2) *no point* $\mathbf{x}$ *of* M *is mapped into its antipode* $-\mathbf{x}$ *by* $\mathbf{v}$, *and if* (3) *the euclidean inner product* $\langle \mathbf{v}(\mathbf{x}), \mathbf{n}(\mathbf{x}) \rangle$ *is positive for every* $\mathbf{x} \in \partial M$, *then the euler number* $\chi(M)$ *is related to the degree* d *of* $\mathbf{v}$ *by the equality*

$$\chi(M) = 1 + (-1)^k d \, .$$

Proof: After perturbing $\mathbf{v}$ slightly we may assume that the fixed points of $\mathbf{v}$ are all isolated. According to the Lefschetz fixed point theorem one can assign an index $\iota(\mathbf{x})$ to each fixed point so that the sum of the indices is equal to the Lefschetz number

$$\sum (-1)^j \text{ Trace } (\mathbf{v}_*: \; H_j(S^k) \to H_j(S^k)) = 1 + (-1)^k d \, .$$

(Compare ALEXANDROFF and HOPF.)

Consider the one-parameter family of mappings

$$v_t : M \to S^k$$

defined by

$$v_t(x) = ((1-t)x + tv(x))/\|(1-t)x + tv(x)\| .$$

This formula makes sense since $v(x) \neq -x$ for $x \in M$. Clearly v_0 is the identity and v_t maps M into itself, by a mapping homotopic to the identity, for small values of t. So the Lefschetz number of

$$v_t : M \to M$$

must be equal to the euler number $\chi(M)$, say for $0 < t \leq \varepsilon$.

But the fixed points of v_t are precisely the same as the fixed points of $v : S^k \to S^k$, for $t > 0$. Since the Lefschetz index of the fixed point x of v_t is an integer which varies continuously with t, it follows that the Lefschetz number $\chi(M)$ of v_ε must be equal to the Lefschetz number $1 + (-1)^k d$ of v. This proves Lemma 7.4.

Proof of Theorem 7.2. Let M be the region consisting of all points $z \in S_\varepsilon$ which satisfy the inequality

$$\Re f(z) \geq 0 .$$

In other words, M is the union of the fibers F_θ as θ ranges over the interval $[-\pi/2, \pi/2]$, together with the common boundary K. Clearly

$$\partial M = F_{-\pi/2} \cup K \cup F_{\pi/2}$$

is a smooth manifold. (Compare the proof of 6.1.)

Note that M has the homotopy type of F_θ. In fact the interior of M is fibered over an open semi-circle with F_θ as fiber.

Consider the smooth function

$$v(z) = \varepsilon \operatorname{grad} f(z)/\|\operatorname{grad} f(z)\|$$

from the sphere S_ε to itself. We will show that v satisfies the three hypotheses of Lemma 7.4.

Hypothesis (1). Clearly $\mathbf{z}$ is a fixed point of $\mathbf{v} = \varepsilon\, \mathbf{grad}\, f/\|\mathbf{grad}\, f\|$ if and only if $\mathbf{grad}\, f(\mathbf{z})$ is a positive real multiple of $\mathbf{z}$. But if

$$\mathbf{grad}\, f(\mathbf{z}) = c\mathbf{z}, \qquad c > 0,$$

then $f(\mathbf{z}) \neq 0$ (compare the proof of 6.1), and

$$\mathbf{grad} \log f(\mathbf{z}) = c\mathbf{z}/\overline{f}(\mathbf{z})$$

where the coefficient $c/\overline{f}(\mathbf{z})$ must have positive real part by §4.3. Hence $\Re f(\mathbf{z}) > 0$, and $\mathbf{z}$ is an interior point of M.

Hypothesis (2) is verified by a similar argument.

Hypothesis (3). Given any boundary point $\mathbf{z}$ of M we can choose a smooth path $\mathbf{p}(t)$ crossing into M with velocity vector $d\mathbf{p}/dt = \mathbf{n}(\mathbf{z})$ at $\mathbf{p}(0) = \mathbf{z}$. Clearly the derivative of $\Re f(\mathbf{p}(t))$ is positive, at $t = 0$, by the definition of M. So the identity

$$d\,\Re f/dt = \Re\langle d\mathbf{p}/dt, \mathbf{grad}\, f\rangle$$

shows that the euclidean inner product $\Re\langle \mathbf{n}(\mathbf{z}), \mathbf{v}(\mathbf{z})\rangle$ is positive.

Hence 7.4 applies, and we have the formula

$$(1) \qquad \chi(F_\theta) = \chi(M) = 1 - \text{degree}(\mathbf{v}),$$

since the dimension $2n+1$ of S_ε is odd.

But the degree of the mapping $\mathbf{v}$ is equal to $(-1)^{n+1}$ times the multiplicity μ of the origin as solution to the set of polynomial equations

$$\partial f/\partial z_1 = \cdots = \partial f/\partial z_{n+1} = 0 .$$

For μ was defined as the degree of the mapping

$$\mathbf{z} \mapsto \mathbf{g}(\mathbf{z})/\|\mathbf{g}(\mathbf{z})\|$$

on S_ε where $\mathbf{g}(\mathbf{z})$ is the complex conjugate of $\mathbf{grad}\, f(\mathbf{z})$. And the conjugation map

$$(g_1, \ldots, g_{n+1}) \to (\overline{g}_1, \ldots, \overline{g}_{n+1})$$

clearly carries S_ε into itself with degree $(-1)^{n+1}$.

Substituting this fact into the formula (1) we obtain

$$\chi(F_\theta) = 1 + (-1)^n \mu .$$

But by definition the euler number $\chi(F_\theta)$ is equal to

$$\sum (-1)^j \operatorname{rank} H_j(F_\theta) = 1 + (-1)^n \operatorname{rank} H_n(F_\theta) .$$

Therefore

$$\mu = \operatorname{rank} H_n(F_\theta)$$

which completes the proof.

§8. IS K A TOPOLOGICAL SPHERE?

Again assume that the origin is an isolated critical point of the polynomial $f(z_1, \ldots, z_{n+1})$, with $n \geq 1$. How can we decide whether or not the compact $(2n-1)$-dimensional manifold $K = f^{-1}(0) \cap S_\varepsilon$ is a topological sphere?

LEMMA 8.1. *If* $n \neq 2$ *then* K *is homeomorphic to the sphere* S^{2n-1} *if and only if* K *has the homology of a sphere.*

For if $n \geq 3$ then K is simply connected by §5.2, and has dimension ≥ 5, so we can apply the generalized Poincaré hypothesis, as verified by SMALE and STALLINGS. Since the statement is trivially true for $n = 1$, this completes the proof.

REMARK. For $n = 2$ the corresponding statement is definitely false. In fact MUMFORD has shown that the fundamental group $\pi_1(K)$ can never be trivial in this case. (See also HIRZEBRUCH, *The Topology of Normal Singularities.*) As an example, consider the polynomial

$$f(z_1, z_2, z_3) = z_1^2 + z_2^3 + z_3^5$$

of the type considered by Brieskorn. Hirzebruch points out that the corresponding 3-manifold K is a homology sphere; but that $\pi_1(K)$ is the perfect group with 120 elements, isomorphic to $SL(2, Z_5)$. (Compare §9.8.) This Poincaré manifold K is familiar to knot theorists as the p-fold branched covering of the 3-sphere, branched along a torus knot of type (q, r), where p, q, r is any permutation of 2, 3, 5.

The criterion 8.1 can be sharpened as follows.

LEMMA 8.2. *For* $n \neq 2$ *the manifold* K *is a topological sphere if and only if the reduced homology group* $\tilde{H}_{n-1}K$ *is trivial.*

For if this group is trivial, then, using Poincaré duality and the fact that K is $(n-2)$-connected, we easily verify that K is a homology sphere.

Now choose an orientation for the $2n$-dimensional orientable manifold F_θ and note that any two n-dimensional homology classes α, β of F_θ have a well-defined intersection number $s(\alpha, \beta)$.

LEMMA 8.3. *The manifold* K *is a homology sphere if and only if the intersection pairing*

$$s: H_n F_\theta \otimes H_n F_\theta \to Z$$

has determinant ± 1.

Proof: This follows from the homology exact sequence

$$H_n \bar{F}_\theta \xrightarrow{j_*} H_n(\bar{F}_\theta, K) \xrightarrow{\partial} \tilde{H}_{n-1}K \longrightarrow 0$$

of the pair $(\bar{F}_\theta, K)$, where the first group is known to be free abelian of rank μ. It follows from the Poincaré duality theorem that $H_n(\bar{F}_\theta, K)$ is also free abelian of rank μ and that the intersection pairing

$$s': H_n(\bar{F}_\theta, K) \otimes H_n \bar{F}_\theta \to Z$$

has determinant ± 1. Using the identity

$$s(\alpha, \beta) = s'(j_* \alpha, \beta)$$

it follows that j_* is an isomorphism if and only if s has determinant ± 1. This proves 8.3.

Here is a different approach to the group $\tilde{H}_{n-1}K$. Given any fiber bundle

$$\phi: E \to S^1$$

over the circle, the natural action of a generator of $\pi_1(S^1)$ on the homology of the fiber is described by an automorphism

$$h_*: H_*F_0 \to H_*F_0 \ .$$

Here h denotes the *characteristic homeomorphism* of the fiber $F_0 = \phi^{-1}(1)$. It is obtained, using the covering homotopy theorem, by choosing a continuous one-parameter family of homeomorphisms

$$h_t: \ F_0 \to F_t$$

for $0 \le t \le 2\pi$, where h_0 is the identity and $h = h_{2\pi}$ is the required characteristic homeomorphism.

LEMMA 8.4. (Wang). *To any such fibration there is associated an exact sequence of the form*

$$\cdots \longrightarrow H_{j+1}E \longrightarrow H_jF_0 \xrightarrow{\ h_* - I_*\ } H_jF_0 \longrightarrow H_jE \longrightarrow \cdots ,$$

where I *denotes the identity map of* F_0 *and* h *denotes the characteristic homeomorphism.*

[The proof can be outlined as follows. The covering homotopy $\{h_t\}$ induces a mapping

$$F_0 \times [0, 2\pi] \to E$$

which gives rise to an isomorphism

$$H_j(F_0 \times [0, 2\pi], \ F_0 \times [0] \cup F_0 \times [2\pi]) \xrightarrow{\ \approx\ } H_j(E, F_0)$$

of relative homology groups. Identifying the left hand group with $H_{j-1}F_0$ and substituting this in the exact sequence of the pair (E, F_0), we obtain the required WANG sequence.]

Now let us specialize to the fibration $\phi: S_\varepsilon - K \to S^1$ of §6. Let $\Delta(t)$ denote the characteristic polynomial

$$\Delta(t) = \det(tI_* - h_*)$$

of the linear transformation

$$h_*: H_nF_0 \to H_nF_0 .$$

Thus $\Delta(t)$ is a polynomial with integer coefficients of the form

$$t^\mu + a_1 t^{\mu-1} + \cdots + a_{\mu-1} t \pm 1 .$$

THEOREM 8.5. *For $n \neq 2$ the manifold K is a topological sphere if and only if the integer*

$$\Delta(1) = \det(I_* - h_*)$$

is equal to ± 1.

Proof: For $n > 1$ this follows immediately from the Wang sequence

$$H_nF_0 \xrightarrow{h_* - I_*} H_nF_0 \longrightarrow H_n(S_\varepsilon - K) \longrightarrow 0 ,$$

together with the Alexander duality isomorphism

$$H_n(S_\varepsilon - K) \cong H^nK$$

and the Poincaré duality isomorphism

$$H^nK \cong H_{n-1}K .$$

(Compare 8.2.) For $n = 1$ a similar argument applies.

REMARK 8.6. The polynomial $\Delta(t)$ can also be obtained in a different way. In place of the automorphism h of F_0 we can make use of the covering transformations in the infinite cyclic covering space $\tilde{E}$ of the complement $E = S_\varepsilon - K$. (Compare LEVINE.) It is easily verified that $\tilde{E}$ is homeomorphic to $F_0 \times R$, and that a suitable generator of the group of covering transformations corresponds to the homeomorphism

$$(\mathbf{z}, r) \mapsto (h(\mathbf{z}), r - 2\pi)$$

of $F_0 \times R$. This shows that $\Delta(t)$ is a topological invariant of $S_\varepsilon - K$, at least when K is connected. This invariant represents an n-dimensional generalization of the Alexander polynomial of a knot (Compare §10.1.)

REMARK 8.7. If K does happen to be a topological sphere, it is interesting to ask which differentiable structure it has. Since K bounds the $(n-1)$-connected parallelizable manifold $\bar{F}_0$, the diffeomorphism class of K is completely determined by the signature of the intersection pairing

$$H_nF_0 \otimes H_nF_0 \to Z$$

if n is even, or by the Kervaire invariant

$$c(F_0) \in Z_2$$

if n is odd. (See KERVAIRE and MILNOR, §7.5 and §8.5. The case $n = 2$ must again be excluded.)

When n is odd a remarkable theorem of LEVINE asserts that the Kervaire invariant is given by

$$c(F_0) = 0 \text{ if } \Delta(-1) \equiv \pm 1 \pmod 8,$$

$$c(F_0) = 1 \text{ if } \Delta(-1) \equiv \pm 3 \pmod 8.$$

Thus for n odd the characteristic polynomial $\Delta(t)$ completely determines the differentiable structure of K, at least when K is a topological sphere.

§9. BRIESKORN VARIETIES AND WEIGHTED HOMOGENEOUS POLYNOMIALS

Given integers $a_1, \ldots, a_{n+1} \geq 2$ consider the polynomial

$$f(z_1, \ldots, z_{n+1}) = (z_1)^{a_1} + (z_2)^{a_2} + \cdots + (z_{n+1})^{a_{n+1}} .$$

Clearly the origin is the only critical point of f, so the intersection of $V = f^{-1}(0)$ with S_ε is a smooth manifold K of dimension $2n-1$. Consider the associated fibration $\phi: S_\varepsilon - K \to S^1$ with fibers F_θ of dimension 2n.

THEOREM 9.1. (Brieskorn-Pham). *The group* $H_n F_\theta$ *is free abelian of rank*

$$\mu = (a_1 - 1)(a_2 - 1) \cdots (a_{n+1} - 1) .$$

The characteristic roots of the linear transformation

$$h_*: H_n(F_0 ; C) \to H_n(F_0 ; C)$$

are the products $\omega_1 \omega_2 \cdots \omega_{n+1}$ *where each* ω_j *ranges over all* a_j-th *roots of unity other than* 1. *Hence the characteristic polynomial is given by*

$$\Delta(t) = \Pi\,(t - \omega_1 \omega_2 \cdots \omega_{n+1}) .$$

An alternative expression for $\Delta(t)$ will be given in §9.6.

As an example, consider the generalized trefoil knot $K \subset S_\varepsilon$ corresponding to the choice of exponents

$$a_1 = \cdots = a_n = 2, \quad a_{n+1} = 3 .$$

Then

$$\omega_1 = \cdots \omega_n = -1, \qquad \omega_{n+1} = (-1 \pm \sqrt{-3})/2 ,$$

so that

$$\Delta(t) = t^2 - t + 1 \text{ for } n \text{ odd}$$

$$\Delta(t) = t^2 + t + 1 \text{ for } n \text{ even.}$$

Thus for n odd we have $\Delta(1) = 1$, so that K is a topological sphere of dimension $2n - 1 = 1, 5, 9, 13, \ldots$. If the dimension of K is 1 or 5, then of course K is diffeomorphic to the standard sphere since there are no exotic spheres in these dimensions. But if $2n - 1 = 9$ then the manifold K is diffeomorphic to KERVAIRE'S exotic 9-sphere. (Compare §8.7.)

Similar examples of exotic and non-exotic spheres in the other dimensions $7, 11, 15, \ldots$ have been given by HIRZEBRUCH and BRIESKORN. Every exotic sphere which embeds in codimension 2 can be obtained in this way.

Theorem 9.1 raises several question. Is there an algorithm for computing the characteristic polynomial $\Delta(t)$ associated with an arbitrary isolated critical point? Is $\Delta(t)$ always a product of cyclotomic polynomials ?* Can the structural group of the fiber bundle always be reduced to a finite group (or at least to a compact group)?

(In the classical case $n = 1$ the Alexander polynomial has been computed by ZARISKI when the knot K is connected, and by BURAU when K has at most two components. In these cases $\Delta(t)$ is always a product of cyclotomic polynomials.)

Proof of Theorem 9.1. It is convenient to set $m = n + 1$.

Note first (using the special form of the polynomial f) that the fibration $\phi: S_\varepsilon - K \to S^1$ extends to a locally trivial fibration

$$\psi: C^m - V \to S^1 ,$$

where ψ, like ϕ, is defined by the formula

$$\psi(\mathbf{z}) = f(\mathbf{z})/|f(\mathbf{z})| .$$

* Addendum: A proof that $\Delta(t)$ is necessarily a product of cyclotomic polynomials has recently been given by A. Grothendieck. Related results were described by Grothendieck in the S.G.A. Seminar at Bures, 1968.

It is easy to verify that ψ is locally trivial, making use of the one-parameter group of diffeomorphisms

$$h_t : C^m - V \to C^m - V$$

defined by

$$h_t(z_1, \ldots, z_m) = (e^{it/a_1} z_1, \ldots, e^{it/a_m} z_m) \; .$$

Note that h_t carries each fiber $\psi^{-1}(y)$ diffeomorphically onto the fiber $\psi^{-1}(e^{it}y)$. We will be particularly concerned with the characteristic homeomorphism $h_{2\pi}$.

Note also that each fiber $\psi^{-1}(y)$ is diffeomorphic to $\phi^{-1}(y) \times R$ under the correspondence

$$(\mathbf{z}, r) \mapsto (e^{r/a_1} z_1, \ldots, e^{r/a_m} z_m) \; ,$$

for $\mathbf{z} \in S_\varepsilon - K$, $r \in R$. Thus the new fibration has the same fiber homotopy type as the old one.

Let Ω_a denote the finite cyclic group consisting of all a-th roots of unity, and let J denote the join

$$J = \Omega_{a_1} * \Omega_{a_2} * \cdots * \Omega_{a_m} \subset C^m$$

consisting of all linear combinations

$$(t_1\omega_1, t_2\omega_2, \ldots, t_m\omega_m)$$

with

$$t_1 \geq 0, \ldots, t_m \geq 0, \qquad t_1 + \cdots + t_m = 1 \; ,$$

and with $\omega_j \in \Omega_{a_j}$. Note that J is contained in $\psi^{-1}(1)$.

LEMMA 9.2 (Pham). *This join* J *is a deformation retract of the fiber* $\psi^{-1}(1)$.

Proof: Given any point $\mathbf{z} \in \psi^{-1}(1)$, first deform each coordinate z_j along a path in C which is chosen so that the a_j-th power of z_j moves in a straight line to the nearest point $\Re(z_j^{a_j})$ of the real axis. Thus the vector $\mathbf{z}$ moves to a vector $\mathbf{z}'$ which satisfies $(z_j')^{a_j} \in R$ for each j. It is clear that the function value $f(\mathbf{z}) > 0$ does not change during this deformation; so that we remain within the fiber $\psi^{-1}(1)$. Next for each j such that $(z_j')^{a_j} < 0$ move z_j' along a straight line to zero, leaving z_j' fixed if $(z_j')^{a_j} \geq 0$. Thus the vector $\mathbf{z}'$ moves in a straight line to a vector $\mathbf{z}'' \in \psi^{-1}(1)$ which satisfies $(z_j'')^{a_j} \geq 0$ for all j. It follows that each coordinate z_j'' is of the form $t_j\omega_j$ for some $t_j \geq 0$ and some $\omega_j \in \Omega_{a_j}$. Finally move $\mathbf{z}''$ along a straight line to the point

$$\mathbf{z}''/(t_1 + \cdots + t_{n+1}) \in J .$$

Since points of J remain fixed throughout the deformation, this completes the proof of 9.2.

The homology of any join $A * B$ is naturally isomorphic to the direct sum of tensor products:

$$\tilde{H}_{k+1}(A * B) \cong \sum_{i+j=k} \tilde{H}_i A \otimes \tilde{H}_j B ,$$

providing that $\tilde{H}_* A$ has no torsion. (See for example MILNOR, *Universal Bundles, II.*) Since each Ω_{a_j} has homology only in dimension zero, we find inductively that

$$\tilde{H}_{m-1} J = \tilde{H}_0 \Omega_{a_1} \otimes \cdots \otimes \tilde{H}_0 \Omega_{a_m} ,$$

the reduced homology groups of J being trivial in all other dimensions.

Now recall that the characteristic homeomorphism

$$h = h_{2\pi} : \psi^{-1}(1) \to \psi^{-1}(1)$$

is given by the formula

$$h_{2\pi}(\mathbf{z}) = (e^{2\pi i/a_1} z_1, \ldots, e^{2\pi i/a_m} z_m) .$$

Evidently, this homeomorphism carries J into itself, and $h_{2\pi}| J$ can be described as the join

$$r_{a_1} * \cdots * r_{a_m} : J \to J ,$$

where r_a denotes the rotation of Ω_a through an angle of $2\pi/a$ given by the formula

$$r_a(\omega) = e^{2\pi i/a}\omega .$$

Consider the induced homomorphism

$$r_{a*} : \tilde{H}_0(\Omega_a ; C) \to \tilde{H}_0(\Omega_a ; C)$$

of reduced homology groups. The eigenvalues of r_{a*} are clearly the a-th roots of unity, other than 1. [Proof: For each integer ν between 1 and $a-1$ the homology class in $\tilde{H}_0(\Omega_a ; C)$ which associates the coefficient $\omega^\nu \in C$ to each point ω of Ω_a is an eigenvector of r_{a*}, corresponding to the eigenvalue $e^{-2\pi i\nu/a}$.]

Hence the eigenvalues of the tensor product homomorphism

$$(h_{2\pi} | J)_* = r_{a_1 *} \otimes \cdots \otimes r_{a_m *}$$

are the products $\omega_1\omega_2 \cdots \omega_m$, where each ω_j ranges over all a_j-th roots of unity other than 1. This completes the proof of Theorem 9.1.

There is a much larger class of polynomials which are almost as easy to work with as the Brieskorn polynomials. Let $a_1, \ldots, a_m$ be positive rational numbers.

Definition 9.3. The polynomial $f(z_1, \ldots, z_m)$ is *weighted homogeneous* of type $(a_1, \ldots, a_m)$ if it can be expressed as a linear combination of monomials $z_1^{i_1} \cdots z_m^{i_m}$ for which

$$i_1/a_1 + \cdots + i_m/a_m = 1 .$$

This is equivalent to the requirement that

$$f(e^{c/a_1} z_1, \ldots, e^{c/a_m} z_m) = e^c f(z_1, \ldots, z_m)$$

for every complex number c.

LEMMA 9.4. *If the polynomial* f *is weighted homogeneous then the fiber* F *of* §4 *is diffeomorphic to the non-singular hypersurface.*

$$F' = \{\mathbf{z} \in C^m \mid f(\mathbf{z}) = 1\} .$$

As characteristic homeomorphism from F' *(or* F*) to itself one can choose the periodic unitary transformation*

$$h(z_1, \ldots, z_m) = (e^{2\pi i/a_1} z_1, \ldots, e^{2\pi i/a_m} z_m) .$$

The proof is straightforward.

Let $h^j: F' \to F'$ denote the composition of h with itself j times. The fixed point set of h^j is clearly the intersection of F' with a linear subspace L_j of C^m which is defined by equations of the form $z_{i_1} = \cdots = z_{i_k} = 0$. It is not difficult to verify that this fixed point set $F' \cap L_j$ is itself a non-singular hypersurface in L_j.

LEMMA 9.5. *The Lefschetz number of the mapping* $h^j: F' \to F'$ *is equal to the euler number of the fixed point manifold of* h^j.

We will denote this euler (or Lefschetz) number by χ_j.

Proof: First note the more general principal that the Lefschetz number of any isometry of a compact Riemannian manifold is equal to the euler number of its fixed point manifold. (Compare KOBAYASHI.) For the Lefschetz number of any map f of a compact manifold into itself depends only on the behavior of f in a neighborhood of the fixed point set. Hence, in the special case of an isometry, we can first replace the whole manifold by a tubular neighborhood T of the fixed point set, and then apply the Lefschetz formula to the restricted mapping $f \mid T$.

The proof now proceeds as follows. As compact manifold, use the intersection of F' with a large disk D centered at the origin*. Using §2.8 it can be verified that $D \cap F'$ is a deformation retract of F'. Similarly the

* This is of course a manifold with boundary. But, with a little care, the boundary points do not cause any difficulty.

fixed point set $D \cap F' \cap L_j$ is a deformation retract of $F' \cap L_j$. The assertion of Lemma 9.5 now follows easily.

Next consider the Weil zeta function

$$\zeta(t) = \exp \sum_{j=1}^{\infty} \chi_j t^j / j$$

of the mapping h. (Compare MILNOR, *Infinite Cyclic Coverings.*) Since the mapping h is periodic, with period say p, a straightforward computation shows that $\zeta(t)$ can be expressed as a product

$$\zeta(t) = \Pi_{d|p} (1-t^d)^{-r_d}$$

where the exponents $-r_d$ can be computed inductively from the formula

$$\chi_j = \sum_{d|j} d\, r_d \ .$$

(It turns out that r_d is an integer which vanishes unless d divides the period p.)

According to WEIL the zeta function can be expressed as an alternating product of polynomials

$$\zeta(t) = P_0(t)^{-1} P_1(t) P_2(t)^{-1} \cdots P_{m-1}(t)^{\pm 1}$$

where $P_i(t)$ is the determinant of the linear transformation

$$(I_* - th_*): H_i F' \to H_i F' \ .$$

Assume that the origin is an isolated critical point of the polynomial f, so that the fiber F' has homology only in dimensions 0 and $m-1$. Then clearly $P_0(t) = 1-t$, and up to sign $P_{m-1}(t)$ is just the characteristic polynomial $\Delta(t)$ of §8. (This depends on observing that the coefficients of $\Delta(t)$ are symmetric, which is clear since $\Delta(t)$ is the characteristic polynomial of a periodic linear transformation.) Thus we obtain:

THEOREM 9.6. *The euler numbers χ_d of the fixed point manifolds* *of the various iterates* $h^d: F' \to F'$ *are related to the characteristic polynomial* $\Delta(t)$ *by the formula*

$$\Delta(t) = (t-1)^{-1} \Pi_{d|p} (t^d - 1)^{r_d}$$

if m *is odd, or*

$$\Delta(t) = (t-1) \Pi_{d|p} (t^d - 1)^{-r_d}$$

if m *is even; where* $\chi_j = \Sigma_{d|j}\, d\, r_d$.

Here are two examples.

EXAMPLE 9.7. The polynomial

$$z_1{}^2 z_2 + z_2{}^4 = (z_1{}^2 + z_2{}^3) z_2$$

is weighted homogeneous of type (8/3, 4). Setting $u = e^{2\pi i/8}$, the linear transformation

$$h(z_1, z_2) = (u^3 z_1, u^2 z_2)$$

has period 8. Note that h and h^2 have no non-trivial fixed points, but that h^4 has four fixed points on F' (the four solutions to the equations $z_1{}^2 z_2 + z_2{}^4 = 1$, $z_1 = 0$). Computation shows that $\mu = 5$, so the euler number $1-\mu$ of the fixed point set F' of h^8 is equal to -4. Therefore

$$\chi_1 = 0, \quad \chi_2 = 0, \quad \chi_4 = 4, \quad \chi_8 = -4,$$

and

$$r_1 = 0, \quad r_2 = 0, \quad r_4 = 1, \quad r_8 = -1,$$

so that

$$\Delta(t) = (t-1)(t^4-1)^{-1}(t^8-1) = (t-1)(t^4+1) .$$

REMARK. The manifold K, consisting of all zeros of $z_1{}^2 z_2 + z_2{}^4$ on the sphere S_ε, consists of a trefoil knot linked by a circle with linking number equal to 2. The Alexander polynomial of this link is equal to $t_1^3 t_2 + 1$. (Compare §10.1.)

EXAMPLE 9.8. (Compare KLEIN, HIRZEBUCH, CARTAN.) Let G be any finite subgroup of the special unitary group SU(2). Then G operates on the coordinate space C^2 with no fixed points other than the origin. *Assertion*: The ring consisting of all polynomials in two variables which are invariant under the action of G is generated by three polynomials, say p_1, p_2, p_3, which are homogeneous of various degrees. These polynomials are related by a single polynomial equation

$$f(p_1, p_2, p_3) = 0 ,$$

where f is weighted homogeneous. The resulting function $p: C^2 \to C^3$ maps the orbit space C^2/G homeomorphically onto the hypersurface $V = f^{-1}(0) \subset C^3$.

It follows easily that the 3-dimensional manifold S^3/G, with fundamental group isomorphic to G, is mapped homeomorphically onto a submanifold of V which is diffeomorphic to the intersection $K = V \cap S_\varepsilon$.

As an example, if G is the binary icosahedral group (the inverse image of the icosahedral group under the surjection $SU(2) \to SO(3)$), then the polynomials p_1, p_2, p_3 are homogeneous of degrees 30, 20, and 12 respectively. The zeros of p_3, in the projective space consisting of all complex lines through the origin in C^2, form the twelve vertices of a regular icosahedron; the zeros of p_2 form the midpoints of the twenty faces of this icosahedron; and the zeros of p_1 form the midpoints of the thirty edges. These three polynomials are related by the equation

$$p_1^2 + p_2^3 + p_3^5 = 0 .$$

Hence the orbit space C^2/G is isomorphic to the Brieskorn variety V(2, 3, 5), and the Poincaré manifold S^3/G is diffeomorphic to the intersection $K = V(2,3,5) \cap S_\varepsilon$. (Compare §8.)

(It is amusing to note that each non-trivial orbit of G in C^2 can be considered as the set of vertices of a certain regular polytope with six hundred faces, each face being a regular tetrahedron. Compare COXETER *Regular Polytopes,* New York, 1963, Plates IV, VII.)

Similarly if G is cyclic of order k then the lens space S^3/G is diffeomorphic to the manifold $V(2,2,k) \cap S_\varepsilon$; if G is the quaternion group, then $S^3/G \cong V(2,3,3) \cap S_\varepsilon$; and if G is the binary tetrahedral group, then $S^3/G \cong V(2,3,4) \cap S_\varepsilon$.

For the binary octahedral group with 48 elements, the orbit space C^2/G is isomorphic to the variety defined by the weighted homogeneous equation

$$z_1^2 + z_2^3 + z_2 z_3^3 = 0 .$$

Finally, for the binary dihedral group with 4k elements, we obtain the variety

$$z_1^2 + z_2^2 z_3 + z_3^{k+1} = 0 .$$

(In the special case $k = 2$ note that this variety is isomorphic to $V(2,3,3)$.) This completes the description for all of the finite subgroups of $SU(2)$.

REMARK. Note that the weighted homogeneous polynomials which are listed above all have type (a_1, a_2, a_3) satisfying the inequality $1/a_1 + 1/a_2 + 1/a_3 > 1$. If $1/a_1 + 1/a_2 + 1/a_3 \leq 1$, then it is conjectured that the 3-manifold $K = V \cap S_\varepsilon$ has infinite fundamental group, and has an open 3-cell as universal covering space. It is conjectured that this infinite group is nilpotent only if $1/a_1 + 1/a_2 + 1/a_3 = 1$. This occurs for the Brieskorn varieties $V(3,3,3)$, $V(2,4,4)$ and $V(2,3,6)$. (Compare BRIESKORN, *Rationale Singularitäten komplexer Flächen, Inventions math., 4* (1968), 336-358.)

§ 10. THE CLASSICAL CASE: CURVES IN C^2

This section will compare the algebraic geometry associated with a singular point of a complex curve with the corresponding knot theory. (Compare BRAUNER, KÄHLER, ZARISKI, BURAU, REEVE.) In particular it will compare the Alexander polynomial of the link $K = V \cap S_\varepsilon$ with the characteristic polynomial $\Delta(t)$ of §8, and will prove an equality

$$2\delta = \mu + r - 1 \,,$$

which relates the "number of double points" δ at the origin to the multiplicity μ studied in §7 and the number r of branches of V passing through the origin.

Let $f(z_1, z_2)$ be a square-free* polynomial in two complex variables which vanishes at 0. The singular set $\Sigma(V)$ of the curve $V = f^{-1}(0)$ consists of all points in V at which

$$\partial f/\partial z_1 = \partial f/\partial z_2 = 0 \,.$$

(See §2.5.) Since every irreducible constituent of V contains at least one simple point which does not lie on any other irreducible constituent, it follows that $\Sigma(V)$ has dimension less than one, and therefore is finite. So the origin is either a simple point or an isolated singular point, and the results of §§ 6, 7, 8 apply.

Let r be the number of local analytic branches of V passing through the origin. (Compare §3.3.) Then the intersection $K = V \cap S_\varepsilon$ is a smooth compact 1-manifold with r components, or in other words a *link*, in the

* In other words, f should be either irreducible or the product of distinct irreducible polynomials.

3-sphere S_ε. (Compare §2.9, 2.10. A link with one component is called a *knot*.)

LEMMA 10.1. *If* $r = 1$ *then the characteristic polynomial* $\Delta(t)$ *of* §8 *is equal to the Alexander polynomial of the knot* K. *If* $r \geq 2$ *then* $\Delta(t)$ *is related to the Alexander polynomial* $\Delta(t_1, \ldots, t_r)$ *of* K *by the identity*

$$\pm\, t^i \Delta(t) = (t-1)\,\Delta(t, \ldots, t) .$$

(The factor $\pm\, t^i$ must be included since $\Delta(t_1, \ldots, t_r)$ is only well-defined up to multiplication by monomials $\pm\, t_1^{i_1} \cdots t_r^{i_r}$.)

The proof of 10.1 will be given at the end of this section.

Example. The algebraic set

$$z_1^p + z_2^{pq} = 0$$

consists of p non-singular branches, any two of which intersect at the origin with intersection multiplicity q. (Each branch can be defined by a polynomial equation of the form $z_1 = \omega z_2^q$, with $\omega^p = -1$.) The corresponding $K = V \cap S_\varepsilon$ is a torus link consisting of p unknotted circles,

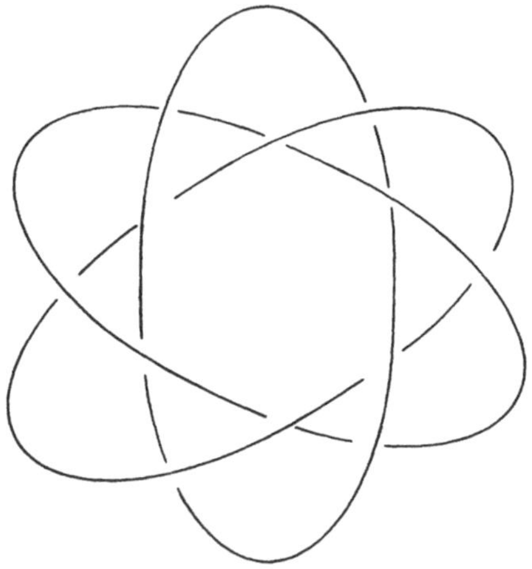

Figure 5. The link K associated with $z_1^3 + z_2^6 = 0$.

any two of which have linking number q. (Compare Figure 5, where the case $p = 3$, $q = 2$ is illustrated.) Computation shows that

$$\Delta(t_1, \ldots, t_p) = ((t_1 \cdots t_p)^q - 1)^{p-1}/(t_1 \cdots t_p - 1)$$

providing that $p \geq 2$. Hence

$$\Delta(t) = (t-1)(t^{pq}-1)^{p-1}/(t^p-1) \ ,$$

with degree μ equal to $(p-1)(pq-1)$. This statement agrees of course with §9.1 and §9.6.

Fibrations of the complement of a knot have been thoroughly studied by STALLINGS and by NEUWIRTH.* They prove the following.

NEUWIRTH-STALLINGS THEOREM. *For a tame knot* k *in the 3-sphere, the following three conditions are equivalent:*

(1) *The complement* $S^3 - k$ *is the total space of a fiber bundle over the circle, the fiber* F *being a connected surface.*

(2) *The commutator subgroup* G' *of the knot group* $G = \pi_1(S^3 - k)$ *is a free group.*

(3) *The commutator subgroup* G' *is a finitely generated group.*

Furthermore, if a knot k *satisfies these conditions, then:*

(4) *the Alexander polynomial of* k *has leading coefficient* ± 1 *and has degree, say* μ, *equal to the rank of the free group* G';

(5) *the fiber* F *is an orientable surface of genus* $\mu/2$ *with just one end***; *and*

(6) *the integer* $\mu/2$ *is equal to the genus of the knot.*

---

* Similar results in higher dimensions have been obtained by BROWDER and LEVINE.

** In other words F can be obtained from a compact surface of genus $\mu/2$ by removing a single point. The proof that F has only one end depends on noting that otherwise a suitably chosen finite cyclic covering of $S^3 - k$ would also have more than one end, which is impossible.

(By definition, the *genus* of a knot k is the minimum genus of an orientable surface spanning k in S^3.)

Applying this result to our fibration we obtain.

COROLLARY 10.2. *If only one branch of the complex curve* V *passes through the origin, then* $K = V \cap S_\varepsilon$ *is a Neuwirth-Stallings knot. The commutator subgroup of* $\pi_1(S_\varepsilon - K)$ *is free of rank* μ, *and the genus of* K *is equal to the genus* $\mu/2$ *of the spanning surface* $\overline{F}_\theta$.

(It follows immediately that this integer μ coincides with the multiplicity μ of §7.2.)

REMARK. Not every Neuwirth-Stallings knot can be obtained in this way. For example, the "figure eight knot" (4_1 in the ALEXANDER-BRIGGS table) is a Neuwirth-Stallings knot, but cannot arise as the knot $V \cap S_\varepsilon$ of a complex singularity since its Alexander polynomial $t^2 - 3t + 1$ is not a product of cyclotomic polynomials. (Compare the discussion in §9.)

Without attempting to prove the Neuwirth-Stallings theorem, let me outline the easy part of the argument, namely the proof that

$$(1) \Longrightarrow (2) \Longrightarrow (3), (4) .$$

If $S^3 - k$ fibers over S^1 with connected fiber F, then the exact sequence

$$\pi_2(S^1) \to \pi_1(F) \to \pi_1(S^3 - k) \to \pi_1(S^1) \to 1$$

shows that $\pi_1(F)$ can be identified with the commutator subgroup G' of $G = \pi_1(S^3 - k)$. Since the fundamental group of any open surface is free, this verifies (2). According to RAPAPORT and CROWELL, the abelianized commutator group G'/G'' of any knot group is torsion free of finite rank.* Hence (2) implies (3). In fact they show that the rank of G'/G'' is equal to the degree of the Alexander polynomial; but that this group is actually finitely generated only if the leading coefficient of the Alexander polynomial is ± 1. Hence (2) implies (4).

* The *rank* of a torsion free abelian group is the maximum number of linearly independent elements.

Now let us look at the algebraic geometry of our singular point. To any singular point $\mathbf{z}$ of a curve $V \subset C^2$ there is associated an integer $\delta_{\mathbf{z}} > 0$ which intuitively measures the number of double points of V concentrated at $\mathbf{z}$. (Compare 10.9.) For a precise definition the reader is referred to SERRE, p. 68.

For our purposes this integer $\delta_{\mathbf{z}}$ can be characterized by two properties.

PROPERTY 10.3. The integer $\delta_{\mathbf{z}}$ is a local analytic invariant. That is, if a complex analytic homeomorphism defined in an open neighborhood of $\mathbf{z}$ carries V locally to some other algebraic curve V' and carries $\mathbf{z}$ to $\mathbf{z}'$, then the integer $\delta_{\mathbf{z}}(V)$ is equal to $\delta_{\mathbf{z}'}(V')$. In fact the same is true if V' is related to V only by a formal power series change of coordinates.

PROPERTY 10.4. If Γ is an irreducible curve of degree d and genus g in the complex projective plane, then

$$\frac{1}{2}(d-1)(d-2) = g + \Sigma\, \delta_{\mathbf{z}} ,$$

to be summed over all singular points $\mathbf{z}$ of Γ.

These two properties are proved by SERRE (pages 68 and 74 respectively).

REMARK. The "genus" of a curve to an algebraic geometer means a certain invariant of the field of rational functions which *a priori* looks very different from the "genus" of a topologist. But a classical theorem asserts that the two definitions coincide in the case of a non-singular complex curve. (See for example SPRINGER or CHEVALLEY.)

THEOREM 10.5. *Suppose that* r *branches of the curve* V *pass through the origin. Then the integer* $\delta = \delta_0(V)$ *is related to the multiplicity* μ *of* §7 *by the equation*

$$2\delta = \mu + r - 1 .$$

For example, when $r = 1$, this says that $2\delta = \mu$, so that δ is equal to the genus of the knot $K = V \cap S_\varepsilon$,

REMARK. The statement of 10.5 is purely algebraic, and should surely hold for curves over other fields of characteristic zero. But the proof will be topological, and applies only to the complex case.

Proof of 10.5: Let $f(z_1, z_2)$ generate the ideal of polynomials which vanish on V. If the degree of $f(z_1, z_2)$ is d, then the corresponding homogeneous equation

$$z_0^d f(z_1/z_0, z_2/z_0) = 0$$

defines a curve $\overline{V}$ in the complex projective plane, the intersection of $\overline{V}$ with the finite plane C^2 being equal to V.

Case 1. Suppose that the completed curve $\overline{V}$ is irreducible and has no singular points other than the original singular point $\mathbf{0} \in V$.

Then the Plücker formula 10.4 reads

$$\frac{1}{2}(d-1)(d-2) = g + \delta\ , \tag{1}$$

where g is the genus of $\overline{V}$ and $\delta = \delta_{\mathbf{0}}(V)$.

Now choose a small constant c and let V_c denote the variety

$$\{(z_1, z_2) | f(z_1, z_2) = c\}\ .$$

It follows from Sard's theorem or Bertini's theorem that V_c has no singular points at all, for almost every choice of c. Clearly the completed variety $\overline{V}_c$ also has no singular points, hence

$$\frac{1}{2}(d-1)(d-2) = g_c \tag{2}$$

where g_c denotes the genus of $\overline{V}_c$. Subtracting (1) from (2) we obtain

$$\delta = g_c - g\ . \tag{3}$$

Now look at the topology of the situation. Recall that V intersects a suitable sphere S_ε transversally in a smooth manifold K consisting of r disjoint circles. If c is sufficiently small, then clearly V_c also intersects S_ε transversally, and the intersection K_c also consists of r disjoint circles.

Recall from §5.11 that the intersection of V_c with the open ε-disk is diffeomorphic with the fiber F_θ. So the manifold-with-boundary $V_c \cap D_\varepsilon$ is connected, with first Betti number equal to μ, and with euler number

$$\chi(V_c \cap D_\varepsilon) = 1 - \mu \ .$$

Since the two manifolds $V_c \cap D_\varepsilon$ and $\bar{V}_c - \operatorname{int} D_\varepsilon$ have union $\bar{V}_c$ and intersection K_c, the euler number $2-2g_c$ of $\bar{V}_c$ must be equal to

$$\chi(V_c \cap D_\varepsilon) + \chi(\bar{V}_c - \operatorname{int} D_\varepsilon) - \chi(K_c) \ .$$

Therefore

$$(4) \qquad 2 - 2g_c = 1 - \mu + \chi(\bar{V}_c - \operatorname{int} D_\varepsilon) \ .$$

Before making a similar computation for the genus g of $\bar{V}$ we must choose a non-singular model, say Γ, for the singular curve $\bar{V}$. Thus Γ is a non-singular projective curve (perhaps in a higher dimensional projective space) and there is a map $\Gamma \to \bar{V}$ which is one-one except that r distinct points of Γ map into the one singular point of $\bar{V}$. Therefore

$$2 - 2g = \chi(\Gamma) = \chi(\bar{V}) + r - 1 \ .$$

Expressing $\bar{V}$ as the union of two subsets $\bar{V} \cap D_\varepsilon$ and $\bar{V} - \operatorname{int} D_\varepsilon$ with intersection K, and noting that $\bar{V} \cap D_\varepsilon$ is contractible by 2.10, we have

$$\chi(\bar{V}) = 1 + \chi(\bar{V} - \operatorname{int} D_\varepsilon) \ ,$$

and therefore

$$(5) \qquad 2 - 2g = \chi(\bar{V} - \operatorname{int} D_\varepsilon) + r \ .$$

But clearly the manifold $(\bar{V} - \operatorname{int} D_\varepsilon)$ is diffeomorphic to $(\bar{V}_c - \operatorname{int} D_\varepsilon)$, if the number c is sufficiently small. Therefore, subtracting (4) from (5) and comparing with (3), we obtain the required formula

$$2\delta = 2(g_c - g) = \mu + r - 1 \ .$$

This proves Case 1 of Theorem 10.5.

Case 2. Suppose that the projective curve $\bar{V}$ is reducible, or has other singular points in addition to the origin.

Then we will modify $f(z_1, z_2)$ by adding to it a homogeneous polynomial

$$h(\mathbf{z}) = c_0 z_1^{\,e} + c_1 z_1^{\,e-1} z_2 + \cdots + c_e z_2^{e}$$

of degree $e \geq d$. Let V' be the curve $f(z_1, z_2) + h(z_1, z_2) = 0$ in C^2, and let $\bar{V}'$ be the corresponding projective curve

$$z_0^e f(z_1/z_0, z_2/z_0) + h(z_1, z_2) = 0 \ .$$

We will prove two lemmas.

LEMMA 10.6. *If the degree* e *of the correction term is sufficiently large, then the integers* δ', μ', r' *associated with the singular point* $\mathbf{0}$ *of* V' *are equal to the corresponding integers* δ, μ, r *associated with the singular point* $\mathbf{0}$ *of* V.

LEMMA 10.7. *If* e *is sufficiently large, then, for almost every choice of the complex coefficients* $c_0, c_1, \ldots, c_e$, *the projective curve* $\bar{V}'$ *will be irreducible, and will have no singular points other than the origin* $z_1 = z_2 = 0$.

Combining these two lemmas, we clearly obtain a proof of Theorem 10.5. For the equation

$$2\delta' = \mu' + r' - 1$$

is true by Case 1 of 10.5, and the equation $2\delta = \mu + r - 1$ certainly follows.

The proofs of 10.6 and 10.7 will be based in turn on the following.

LEMMA 10.8. *Let* f *and* g *be complex** *analytic functions of* m *variables. If* f *has an isolated critical point at* 0 *and if* $g-f$ *vanishes to sufficiently high order at* 0, *then there exists a formal power series of the form*

$$w(\mathbf{z}) = \mathbf{z} + \sum a_{jk} z_i z_j + \text{(higher terms)}$$

so that

$$f(w(\mathbf{z})) = g(\mathbf{z}) \ .$$

REMARK. John Mather has proven the much sharper statement that $w(\mathbf{z})$ can be chosen as a *convergent* power series. (Unpublished.) This would be much more convenient for our purpose, but I will only prove the weaker statement given above.

Proof of 10.8: Since the analytic equations

$$\partial f/\partial z_1 = \cdots = \partial f/\partial z_m = 0$$

define an analytic set with an isolated point at the origin, it follows from the local analytic version of the Nullstellensatz that some power $z_j^{k_j}$ of the j-th coordinate function z_j belongs to the ideal spanned by $\partial f/\partial z_1, \ldots, \partial f/\partial z_m$ in the ring of locally convergent power series. (See GUNNING and ROSSI.) Setting $k = k_1 + \cdots + k_m$, it follows that every monomial $z_1^{i_1} \cdots z_m^{i_m}$ of degree $i_1 + \cdots + i_m \geqq k$ belongs to this ideal.

Now pass to the ring $C[[z_1, \ldots, z_m]]$ of formal power series. Elements of this ring will be denoted by symbols such as $f = f(\mathbf{z})$. Let I be the maximal ideal, spanned by $z_1, \ldots, z_m$. Clearly we have established the following.

ASSERTION. If $e \geq k$, then every element of the ideal I^e can be expressed as a linear combination

---

* The corresponding statement for real analytic functions is false, for example when $f(x_1, x_2) = (x_1^2 + x_2^2)^2$.

$$a_1\, \partial f/\partial z_1 + \cdots + a_m\, \partial f/\partial z_m$$

with coefficients

$$a_1, \ldots, a_m \in I^{e-k} \ .$$

Suppose that $f \equiv g \bmod I^{2k+1}$. Setting

$$g(\mathbf{z}) - f(\mathbf{z}) = \sum a_j^1(\mathbf{z})\, \partial f/\partial z_j$$

with $a_j^1 \in I^{k+1}$, we can form the Taylor expansion

$$f(\mathbf{z} + \mathfrak{a}^1(\mathbf{z})) = f(\mathbf{z}) + \sum a_j^1(\mathbf{z})\, \partial f/\partial z_j$$

$$+ \frac{1}{2} \sum a_i^1(z) a_j^1(z)\, \partial^2 f/\partial z_i \partial z_j + \cdots$$

$$\equiv g(\mathbf{z}) \bmod I^{2k+2} .$$

Suppose by induction that we can find elements

$$a_j^2 \in I^{k+2}, \ldots, a_j^s \in I^{k+s}$$

so that

$$f(\mathbf{z} + \mathfrak{a}^1(\mathbf{z}) + \cdots + \mathfrak{a}^s(\mathbf{z})) \equiv g(\mathbf{z}) \bmod I^{2k+s+1} \ .$$

Denoting the left side of this equation by $f'(\mathbf{z})$, a similar argument constructs elements $b_j \in I^{k+s+1}$ so that

$$f'(\mathbf{z} + \mathbf{b}(\mathbf{z})) \equiv g(\mathbf{z}) \bmod I^{2k+2s+2} \ .$$

Setting

$$\mathfrak{a}^{s+1}(\mathbf{z}) = \mathbf{b}(\mathbf{z}) + \sum_{\nu=1}^{s} (\mathfrak{a}^{\nu}(\mathbf{z}+\mathbf{b}(\mathbf{z})) - \mathfrak{a}^{\nu}(\mathbf{z})) \ ,$$

it follows that

$$f(\mathbf{z}+\mathfrak{a}^1(\mathbf{z}) + \cdots + \mathfrak{a}^{s+1}(\mathbf{z})) = f((\mathbf{z}+\mathbf{b}(\mathbf{z})) + \mathfrak{a}^1(\mathbf{z}+\mathbf{b}(\mathbf{z})) + \cdots + \mathfrak{a}^s(\mathbf{z}+\mathbf{b}(\mathbf{z})))$$

$$= f'(\mathbf{z}+\mathbf{b}(\mathbf{z})) \equiv g(\mathbf{z}) \bmod I^{2k+s+2} \ .$$

Since $a_j^{s+1}(\mathbf{z}) \in I^{k+s+1}$, this completes the induction.

Now, passing to the limit as $s \to \infty$, we obtain the required equation

$$f(\mathbf{z}+\mathfrak{a}^1(\mathbf{z}) + \mathfrak{a}^2(\mathbf{z}) + \cdots) = g(\mathbf{z}) .$$

This completes the proof of Lemma 10.8.

Proof of Lemma 10.6. The equation $\delta' = \delta$ is an immediate consequence of 10.8 together with 10.3. The equation $r' = r$ also follows from 10.8, since "branches" of an algebraic curve can be defined in terms of formal power series parametrizations. (See for example VAN DER WAERDEN *Algebraische Geometrie,* p. 52.)

The equality $\mu' = \mu$ will be proved by a rather different argument. We will work with a function of m variables, since the proof is no more difficult. Since the real polynomial function $\|\mathbf{grad}\, f\|^2$ has an isolated zero at the origin, an inequality of HÖRMANDER and LOJASIEWICZ implies that this function is bounded away from zero by a power of $\|\mathbf{z}\|$, say

$$\|\mathbf{grad}\, f(\mathbf{z})\| \geq c\,\|\mathbf{z}\|^r > 0$$

for $0 < \|\mathbf{z}\| \leq \varepsilon$. (This inequality can also be derived from the local analytic Nullstellensatz.)

Now if the homogeneous polynomial h has degree $\geq r + 2$, then

$$\|\mathbf{grad}\, h(\mathbf{z})\| < c\|\mathbf{z}\|^r$$

for small $\mathbf{z}$. Using Rouché's Principle it follows easily that the degree μ' of the mapping

$$(\partial(f+h)/\partial z_1, \ldots, \partial(f+h)/\partial z_m)/\|(\partial(f+h)/\partial z_1, \ldots, \partial(f+h)/\partial z_m)\|$$

on the sphere S_ε is equal to the degree μ of the mapping

$$(\partial f/\partial z_1, \ldots, \partial f/\partial z_m)/\|(\partial f/\partial z_1, \ldots, \partial f/\partial z_m)\|$$

on S_ε. (Compare Appendix B.) This completes the proof of Lemma 10.6.

Proof of Lemma 10.7: If the degree e of the homogeneous correction term is $\geq d$, then it follows immediately from Bertini's theorem that the modified curve $\overline{V}'$ has no singular points other than the origin, for almost every choice of coefficients. (Compare VAN DER WAERDEN, *Algebraische Geometrie,* p. 201.)

Suppose that the modified polynomial $f(z_1, z_2) + h(z_1, z_2)$ is reducible, splitting as the product of two factors with degrees d_1 and d_2 adding up to e. By Bezout's theorem the two corresponding projective curves must intersect, with total intersection multiplicity equal to $d_1 d_2$.

These curves can intersect only at the origin since their union $\overline{V}'$ has no singular points other than 0. But intersection multiplicities at 0 are clearly invariant under a formal power series change of coordinates. So it follows from 10.8 that the r branches of V through 0 can be partitioned into two subsets with intersection multiplicity equal to $d_1 d_2 \geq e-1$. For e sufficiently large this is clearly impossible, so the polynomial $f(z_1, z_2) + h(z_1, z_2)$ must be irreducible. This completes the proof of Lemma 10.7 and Theorem 10.5.

REMARK 10.9. It would be nice to have a better topological interpretation of the integer δ. It can be shown that the link $K = V \cap S_\varepsilon$ bounds a collection of r smooth 2-cells in the disk D_ε having no singularities other than δ ordinary double points. *Question:* Is δ perhaps equal to the "Überschneidungszahl" of K: the smallest number of times which K must be allowed to cross itself during a smooth deformation so as to transform K into a collection of r unlinked and unknotted circles? (Compare WENDT.)

REMARK 10.10. Here is an explicit formula for δ in the case $r = 1$. Describe the curve V locally in terms of a parameter w by the power series

$$z_1 = w^{a_0}$$
$$z_2 = \lambda_1 w^{a_1} + \lambda_2 w^{a_2} + \lambda_3 w^{a_3} + \cdots ,$$

where the exponents a_j are positive integers with greatest common divisor one and with $a_1 < a_2 < a_3 < \cdots$, and where the coefficients λ_j are non-zero. (Compare §3.3.) Let D_j denote the greatest common divisor of $\{a_0, a_1, \ldots, a_{j-1}\}$ so that $a_0 = D_1 \geq D_2 \geq \cdots \geq D_k = 1$ for large k. Then

$$\mu = 2\delta = \sum_{j \geq 1} (a_j - 1)(D_j - D_{j+1}) \; .$$

The proof will be omitted.

In the case $r > 1$ the integer δ can be described as a sum

$$\delta = \delta_{(1)} + \cdots + \delta_{(r)} + \sum_{i<j} \delta_{ij}$$

where $\delta_{(i)} \geq 0$ denotes the integer δ associated with the i-th branch, and $\delta_{ij} > 0$ denotes the intersection multiplicity between the i-th branch and the j-th branch. (Thus $\delta \geq r(r-1)/2$, and hence $\mu = 2\delta - r + 1 \geq (r-1)^2$.)

REMARK 10.11. Here is a somewhat more explicit description of the knot K, following BURAU and KÄHLER. The parametrization above can always be chosen so that $a_0 < a_1$. Then K is a "compound cable knot" which can be constructed, starting with an unknotted circle k_0, by choosing a knot k_1 in the boundary of a tubular neighborhood of k_0, then choosing k_2 in the boundary of a tubular neighborhood of k_1, and so on; the construction being iterated as many times as there are non-zero summands in the above formula for $\mu = 2\delta$.

To conclude this section, here is a proof of 10.1. First recall the definition of the ALEXANDER polynomial of a link (due to FOX when $r \geq 2$).

Given any group G, let X be a connected complex with fundamental group G. Any normal subgroup $N \subset G$ determines a regular covering $\tilde{X}$ with $\pi_1(\tilde{X}) = N$. Let X^0 be a base point in X and let $\tilde{X}^0$ be its full inverse image in $\tilde{X}$. Then the homology $H_1(\tilde{X}, \tilde{X}^0)$ can be thought of as a module over the integral group ring $Z[G/N]$ of the group of covering transformations. The isomorphism class of this module depends only on G and N.

In particular, given a presentation of G with p generators and q relations, we can choose X to be a 2-dimensional complex with a single vertex, with a 1-cell corresponding to each generator, and with a 2-cell corresponding to each relation. The exact sequence

$$H_2(\tilde{X}, \tilde{X}^1) \to H_1(\tilde{X}^1, \tilde{X}^0) \to H_1(\tilde{X}, \tilde{X}^0) \to 0 ,$$

where the first two modules are free, then shows that the module $H_1(\tilde{X}, \tilde{X}^0)$ has a presentation with p generators and q relations. (Here $\tilde{X}^1$ denotes the 1-skeleton.)

If the group G/N of covering transformations is commutative, then the $(p-i) \times (p-i)$ minor determinants of a corresponding relation matrix span an ideal $\mathcal{E}_i^N \subset Z[G/N]$ which is an invariant of the module $H_1(\tilde{X}, \tilde{X}^0)$. (Compare ZASSENHAUS.) In particular, if G' is the commutator subgroup of G then the ideals $\mathcal{E}_i^{G'}$ are certainly defined.

If G is the group of a knot and N is the commutator subgroup G', then the "order ideal" $\mathcal{E}_0^{G'}$ turns out to be zero; and $\mathcal{E}_1^{G'}$ is a principal ideal. A generator of $\mathcal{E}_1^{G'}$ is called the *Alexander polynomial* of the knot.

If G is the group of a link with $r \geq 2$ oriented components, then G/G' is free abelian with generators $t_1, \ldots, t_r$ corresponding to the various components. In this case also $\mathcal{E}_0^{G'} = 0$; and the ideal $\mathcal{E}_1^{G'}$ is equal to the fundamental ideal $(t_1 - 1, \ldots, t_r - 1)$ multiplied by a principal ideal. Again a generator of the principal ideal is called the *Alexander polynomial.*

To prove 10.1 we must study the infinite cyclic covering $\tilde{E}$ of $E = S_\varepsilon - K$ which arises from the universal covering of the base space S^1. This covering corresponds to a certain normal subgroup N of $G = \pi_1(S_\varepsilon - K)$. Clearly the natural homomorphism

$$Z[G/G'] \to Z[G/N]$$

maps all of the generators $t_1, \ldots, t_r$ of G/G' to the single generator, say t, of G/N. Evidently the ideals

$$\mathcal{E}_i^N \subset Z[G/N]$$

are the images under this homomorphism of the corresponding ideals $\mathcal{E}_i^{G'}$. In particular, assuming that $r \geq 2$, the ideal

$$\mathcal{E}_1^{G'} = (t_1 - 1, \ldots, t_r - 1)\,\Delta(t_1, \ldots, t_r)$$

must map onto $\mathcal{E}_1^N$, hence

$$\mathcal{E}_1^N = ((t-1)\,\Delta(t, \ldots, t)) .$$

Now consider the exact sequence

$$0 \longrightarrow H_1\tilde{E} \longrightarrow H_1(\tilde{E}, \tilde{E}^0) \xrightarrow{\partial} H_0\tilde{E}^0 \longrightarrow H_0\tilde{E} \longrightarrow 0 .$$

(Compare CROWELL, "Corresponding group and module sequences.") It is easily verified that $H_0\tilde{E}^0$ is a free $Z[G/N]$ module on one generator, say ξ, and that $\partial H_1(\tilde{E}, \tilde{E}^0)$ is the free submodule generated by $(t-1)\xi$. Hence

$$H_1(\tilde{E}, \tilde{E}^0) \cong H_1\tilde{E} \oplus Z[G/N] .$$

So the first elementary ideal

$$\mathcal{E}_1^N(H_1(\tilde{E}, \tilde{E}^0)) = ((t-1)\,\Delta(t, \ldots, t))$$

is clearly equal to the order ideal $\mathcal{E}_0^N(H_1\tilde{E})$.

But this order ideal is clearly spanned by the characteristic polynomial $\Delta(t)$ of the linear transformation which carries each element a of the free abelian group $H_1\tilde{E} \cong H_1F_\theta$ onto $t_*(a)$. (Compare §8.6.) Thus we obtain the required formula

$$((t-1)\,\Delta(t, \ldots, t)) = (\Delta(t)) ,$$

in the case $r \geq 2$.

Since the discussion for $r = 1$ is completely analogous, this completes the proof.

§11. A FIBRATION THEOREM FOR REAL SINGULARITIES

Let $f: R^m \to R^k$ be a polynomial mapping which takes the origin to the origin, and satisfies the following.

HYPOTHESIS 11.1. *There should exist a neighborhood* U *of the origin in* R^m *so that the matrix* $(\partial f_i/\partial x_j)$ *has rank* k *for all* x *in* U *other than* $x = 0$.

It follows that the equations

$$f_1(x) = \cdots = f_k(x) = 0$$

define an algebraic set V which is a smooth manifold of dimension $m-k$ throughout $U \cap V - \{0\}$. Furthermore the intersection $K = V \cap S_\varepsilon^{m-1}$ is a smooth manifold of dimension $m-k-1$, for small ε. (Compare §2.9. Note that K may be vacuous.)

Assume also that $k \geq 2$.

THEOREM 11.2. *The complement of an open tubular neighborhood of* K *in* S_ε^{m-1} *is the total space of a smooth fiber bundle over the sphere* S^{k-1}, *each fiber* F *being a smooth compact* $(m-k)$*-dimensional manifold bounded by a copy of* K.

Proof: Using 2.9 or 3.1 one can verify that the origin is a regular value of the mapping

$$f \mid S_\varepsilon^{m-1} : S_\varepsilon^{m-1} \to R^k .$$

Hence there exists a small disk D_η^k consisting exclusively of regular values. It follows that the inverse image

$$T = \{x \in S_\varepsilon^{m-1} \mid \|f(x)\| \le \eta\}$$

is fibered over D_η^k with typical fiber K. (Compare EHRESMANN.) Since the base space is contractible, it follows that T is diffeomorphic to the product $K \times D_\eta^k$. We will refer to T as a *tubular neighborhood* of K in S_ε^{m-1}.

Now consider the set $E = D_\varepsilon^m \cap f^{-1}(S_\eta^{k-1})$. (Compare Figure 6.) Note that E is a smooth manifold with $\partial E = \partial T$. An argument similar to that of Ehresmann shows that

$$f|E : E \to S_\eta^{k-1}$$

is also the projection map of a smooth fiber bundle. A typical fiber,

$$F_y = D_\varepsilon^m \cap f^{-1}(y) ,$$

is a compact manifold bounded by the set

$$\partial F_y = S_\varepsilon^{m-1} \cap f^{-1}(y)$$

which is diffeomorphic to K (since it is a fiber of the fibration $T \to D_\eta^k$).

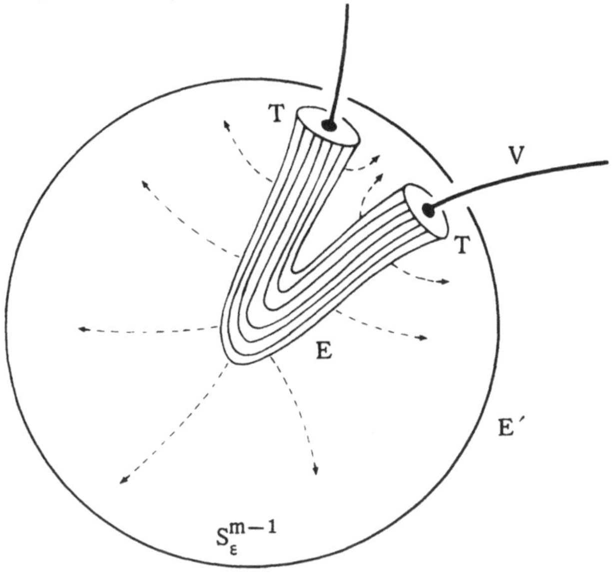

Figure 6.

LEMMA 11.3. *The total space* $E = D_\varepsilon^m \cap f^{-1}(S_\eta^{k-1})$ *of this bundle is diffeomorphic to the complement* $(S_\varepsilon^{m-1} - \text{int } T)$ *of the open tubular neighborhood.*

In other words E is diffeomorphic to the manifold E′ consisting of all $x \in S_\varepsilon^{m-1}$ with $\|f(x)\| \geq \eta$. The proof is similar to that of 5.10. One first needs to construct a vector field $v(x)$ on $D_\varepsilon^m - V$ so that the euclidean inner products $\langle v(x), x\rangle$ and $\langle v(x), \text{grad } \|f(x)\|^2\rangle$ are both positive. This is possible since the two vector fields

$$\text{grad } \|f(x)\|^2$$

and

$$2x = \text{grad } \|x\|^2$$

are non-zero throughout $D_\varepsilon^m - V$, and cannot point in opposite directions by §3.4.

Now pushing out along the trajectories of this vector field v, we map E diffeomorphically onto E′. This proves Lemma 11.3.

Thus E′, the complement of an open tubular neighborhood of K in S_ε^{m-1}, can also be fibered over S_η^{k-1}. This completes the proof of 11.2.

REMARKS. With a little more effort one can prove that the entire complement $S_\varepsilon^{m-1} - K$ also fibers over S^{k-1}, each fiber being the interior of a compact manifold bounded by K.

However it is not true that the obvious mapping

$$x \mapsto f(x)/\|f(x)\|$$

from $S_\varepsilon^{m-1} - K$ to S^{k-1} is the projection map of a fibration. [This direct construction breaks down, for example, in the case

$$f(x_1, x_2) = (x_1, x_1^2 + x_2(x_1^2 + x_2^2)) \text{ .]}$$

Note that any polynomial mapping $R^m \to R^k$ satisfying 11.1 can be composed with the projection $R^k \to R^{k-1}$ to obtain a new mapping

$R^m \to R^{k-1}$ which certainly also satisfies 11.1. *Conjecture:* The fiber of the fibration associated with this new mapping is homeomorphic to the product of the old fiber with the unit interval.

The major weakness of Theorem 11.2 is that the hypothesis is so strong that examples are very difficult to find.

Problem. For which dimensions $m \geq k \geq 2$ do non-trivial examples exist?

It is not quite clear what "non-trivial" should mean here. Certainly the projection $f(x_1, \ldots, x_m) = (x_1, \ldots, x_k)$ is a trivial example. Here is a tentative definition: An example will be called *trivial* if and only if the fiber F of the fibration $E' \to S^{k-1}$ is diffeomorphic to the disk D^{m-k}. (This implies that K, which is isotopic to ∂F in S_ε^{m-1}, must be an unknotted sphere.)

There are many non-trivial examples with $k = 2$. In fact all of the fibrations of §6 occur as examples: Every complex polynomial $f(z_1, \ldots, z_m)$ with an isolated critical point at the origin gives rise to a polynomial mapping $R^{2m} \to R^2$ which clearly satisfies the hypothesis 11.1.

Problem. Are there other, essentially different, examples when $k = 2$? For example can the figure-eight knot occur as the intersection $V \cap S_\varepsilon$ associated with a polynomial mapping $R^4 \to R^2$? Are there non-trivial examples with m odd and $k = 2$?

If $m < 2(k-1)$ it may be conjectured that all examples are trivial. (In contrast we will exhibit non-trivial examples, due to Kuiper, with $m = 2(k-1) = 4, 8, 16$.)

LEMMA 11.4. *If* $m < 2(k-1)$ *then the fiber* F *is necessarily contractible.*

Proof: The sphere S_ε^{m-1} can be obtained from the subspace $E' = S_\varepsilon^{m-1} - \operatorname{int} T$ by adjoining a number of cells of dimension $\geq k$, one $(k+i)$-cell for each i-cell of K. Hence

$$\pi_i E' \cong \pi_i S_\varepsilon^{m-1} = 0$$

for $i \leq k-2$.

Since the fibration $E' \to S^{k-1}$ has a cross section $S^{k-1} \to \partial E' \subset E'$, it follows that the sequence

$$0 \to \pi_i F \to \pi_i E' \to \pi_i S^{k-1} \to 0$$

is split exact. Hence the fiber F is also $(k-2)$-connected.

Now suppose that $m < 2(k-1)$. Then $k \geq 3$, hence F is simply connected. As in §6.2 there is an Alexander duality isomorphism

$$\tilde{H}_i F \cong \tilde{H}^{m-i-2} F .$$

So if the dimension $m-k$ of F is less than half of $m-2$, it follows that F has the homology of a point. And a simply connected space with the homology of a point is certainly contractible.

Since the condition $m-k < \frac{1}{2}(m-2)$ is equivalent to the hypothesis $m < 2(k-1)$ of Lemma 11.4, this completes the proof.

Thus if $m < 2(k-1)$ then F is contractible, and it follows easily that K is a homology sphere. If the dimension $m-k$ of F is ≤ 2, then F is an actual cell, and the example is a "trivial" one. But if $m-k \geq 3$, I cannot prove that F is a cell; and if $m-k \geq 4$, I cannot prove that K is simply connected.

Next I will show that K cannot be an exotic or knotted sphere providing that the codimension k is ≥ 3, and providing that $m-k \geq 6$.

LEMMA 11.5. *If the codimension* k *is* ≥ 3, *and if* K *has the homology of a sphere, then the fiber* F *must be contractible.*

So if K is actually a homotopy sphere of dimension $m-k-1 \geq 5$, then it follows from SMALE that F is diffeomorphic to a disk, and hence that the example is "trivial." I do not know whether this is true for $m-k-1 = 2, 3, 4$.

Proof: As in the proof of 11.4, the fiber F is $(k-2)$-connected. Since K is a homology $(m-k-1)$-sphere it follows by Alexander duality that the space E' is a homology $(k-1)$-sphere. Using the WANG sequence

$$\cdots \to H_1F \to H_{k-1}F \to H_{k-1}E' \to H_0F \to H_{k-2}F \to \cdots$$

of the fibration, it then follows easily that F has the homology of a point; which completes the proof.

To conclude this section let me describe some examples suggested by N. Kuiper. First I will show that the Hopf fibrations $S^{2p-1} \to S^p$, $p = 2$, 4, 8, can be obtained from Theorem 11.2. (Compare P. Baum, Illinois J. Math. 11, p. 586.)

Let A denote either the complex numbers, the quaternions, or the Cayley numbers. Define

$$f: A \times A \to A \times R$$

by

$$f(x, y) = (2x\bar{y}, |y|^2 - |x|^2) .$$

LEMMA 11.6. *This mapping* f *carries the unit sphere of* $A \times A$ *to the unit sphere of* $A \times R$ *by a Hopf fibration.*

Proof: The identity $\|(f(x, y)\|^2 = |2x\bar{y}|^2 + (|y|^2 - |x|^2)^2 = (|x|^2 + |y|^2)^2$ shows that f carries the unit sphere of $A \times A$ to the unit sphere of $A \times R$. Now follow f by the stereographic projection

$$\sigma(z, t) = z/(1 + t)$$

which carries the unit sphere of $A \times R$ with the point $(0, -1)$ removed diffeomorphically onto A. Then

$$\begin{aligned}\sigma f(x, y) &= 2x\bar{y}/(1 + |y|^2 - |x|^2) \\ &= 2x\bar{y}/2|y|^2 = xy^{-1}\end{aligned}$$

(assuming that $|x|^2 + |y|^2 = 1$).

Comparing this formula with the explicit definition of the Hopf fibration (compare STEENROD, p. 109), we see that f restricted to the unit sphere is indeed a Hopf fibration.

Now I can describe the Kuiper examples. With A as above, define

$$f: A^n \times A^n \to A \times R$$

by

$$f(x, y) = (2\langle x, y\rangle, \|y\|^2 - \|x\|^2) ,$$

using the hermitian inner product on A^n. In order to verify that the matrix of (real) first derivatives has maximal rank at every point other than $(0, 0)$, it is clearly sufficient to consider the case $n = 1$. But for $n = 1$ this statement follows easily from 11.6, and the fact that f is homogeneous (of degree 2) over the real numbers.

Thus f satisfies the hypothesis 11.1. Note that f maps a vector space of real dimension either 4n, 8n, or 16n to a vector space of dimension 3, 5, or 9 respectively. The base space of the corresponding fibration is the sphere of dimension 2, 4, or 8 respectively.

The manifold K in this Kuiper example is the Steifel manifold of 2-frames in A^n, and the fiber F is diffeomorphic to a disk bundle over the unit sphere of A^n, consisting of all pairs (x, y) in $A^n \times A^n$ with

$$\|x\| = 1, \quad \|y\| \leq 1, \quad \text{and} \quad \langle x, y\rangle = 0 .$$

APPENDIX A

WHITNEY'S FINITENESS THEOREM FOR ALGEBRAIC SETS

This appendix will present a proof, only slightly different from Whitney's, of Theorem 2.4, which asserted that any difference $V - W$ of real or complex algebraic sets has at most a finite number of topological components.

It is sufficient to consider the real case, since any complex algebraic set in C^m can be thought of as a real algebraic set in R^{2m}.

(REMARK: If V is complex and irreducible, one can actually make the sharper statement that $V - W$ is connected. Compare LEFSCHETZ, *Algebraic Geometry*, p. 97.)

The proof will be based on the following. Let V be an algebraic set in the m-dimensional coordinate space over any infinite field, and let $f_1, \ldots, f_m$ be polynomials which vanish throughout V.

LEMMA A.1. *If the matrix* $(\partial f_i / \partial x_j)$ *is non-singular at a point* x^0 *of* V, *then, removing* x^0 *from* V, *the complement* $V - \{x^0\}$ *will still be an algebraic set.*

In the real case it follows of course that x^0 is an isolated point of V. (But the converse is false: compare Example 2 of §2.)

Proof: We may assume that $x^0 = 0$. Since the polynomial f_j vanishes at the origin, it is easy to choose polynomials g_{jk} so that

$$f_j(x) = g_{j1}(x)\, x_1 + \cdots + g_{jm}(x)\, x_m$$

Let W denote the algebraic set consisting of all points $x \in V$ which satisby the polynomial equation

$$\det(g_{jk}(x)) = 0 .$$

Then the origin is not a point of W, since the matrix

$$(\partial f_j(0)/\partial x_k) = (g_{jk}(0))$$

is non-singular. But at any point $x \neq 0$ of V the linear dependence relation

$$\begin{pmatrix} 0 \\ \vdots \\ 0 \end{pmatrix} = \begin{pmatrix} g_{11}(x) \\ \vdots \\ g_{m1}(x) \end{pmatrix} x_1 + \cdots + \begin{pmatrix} g_{1m}(x) \\ \vdots \\ g_{mm}(x) \end{pmatrix} x_m$$

shows that $\det(g_{jk}(x)) = 0$. So $V - \{0\} = W$, which proves that $V - \{0\}$ is an algebraic set.

Now let us specialize to the field R of real numbers.

COROLLARY A.2. *If an algebraic set* $V \subset R^m$ *has topological dimension zero (for example, if* V *consists only of isolated points), then* V *is a finite set.*

Proof: Let $f_1, \ldots, f_k$ span the ideal I(V). It is enough to show that every zero-dimensional algebraic set V contains at least one point x^0 at which the matrix $(\partial f_i/\partial x_j)$ has rank m. For then the point x^0 can be removed by A.1, yielding a proper algebraic subset $V_1 = V - \{x^0\}$. Iterating this construction, we will obtain a chain

$$V_1 \supset V_2 \supset V_3 \supset \cdots$$

of nested algebraic subsets. Since every such chain must terminate by §2.1, this will prove that V is finite.

But if the matrix $(\partial f_i/\partial x_j)$ had rank at most $\rho \leq m-1$ at all points of V, then Theorem 2.3 would imply that V contained a smooth manifold

$V - \Sigma(V)$ of dimension $m - \rho \geq 1$. Since this would contradict the hypothesis that V has topological dimension zero, this completes the proof of A.2.

LEMMA A.3. *Any non-singular algebraic set* $V \subset R^m$ *has the homotopy type of a finite complex.*

Proof: Given any point $a \in R^m$ let

$$r_a : V \to R$$

denote the squared distance function

$$r_a(x) = \|x - a\|^2 .$$

A lemma of ANDREOTTI and FRANKEL asserts that, for almost every choice of a, the function r_a on V has only non-degenerate critical points.

Let $\Gamma \subset V$ denote the set of all critical points of r_a. According to Lemma 2.7, Γ is an algebraic set. But non-degenerate critical points are clearly isolated, so it follows from A.2 that Γ is a finite set.

An elementary argument now shows that V has only finitely many components. Every component $V^{(i)}$ of V must intersect the critical set Γ. For the distance from a must be minimized at some point x of the closed set $V^{(i)}$, and clearly this closest point x will belong to Γ. Therefore V can have only a finite number of components.

Alternatively, recall the main theorem of Morse theory which states that the manifold V has the homotopy type of a cell complex with one cell for each critical point of the non-degenerate, proper, non-negative function r_a. (Compare MILNOR, *Morse Theory,* §3.5 as well as §6.6.) Then the finiteness of Γ implies the much sharper statement that V has the homotopy type of a finite complex. This proves A.3.

COROLLARY A.4. *For any real algebraic set* V, *if* W *is an algebraic subset containing the singular set* $\Sigma(V)$, *then* $V - W$ *has the homotopy type of a finite complex.*

Proof: Suppose that W is defined by polynomial equations $f_1(x) = \cdots = f_k(x) = 0$. Setting

$$s(x) = f_1(x)^2 + \cdots + f_k(x)^2 ,$$

note that W can also be defined* by the single polynomial equation $s(x) = 0$.

Now let G be the graph of the rational function $1/s$ from V to R. That is, let G be the set of all

$$(x, y) \in V \times R \subset R^{m+1}$$

for which $s(x)y = 1$.

Then clearly G is an algebraic set, and is homeomorphic to $V - W$. Since an easy computation shows that G has no singular points, this proves A.4.

THEOREM. *For any pair* $V \supset W$ *of real algebraic sets, the difference* $V - W$ *has at most a finite number of path-components.*

Proof: According to §2.5 the set V can be expressed as a finite union $M_1 \cup \cdots \cup M_p$, where the manifold M_1 is the set of non-singular points of $V_1 = V$, the manifold M_2 is the set of non-singular points of $V_2 = \Sigma(V_1)$, and so on. Therefore

$$V - W = (M_1 - W) \cup \cdots \cup (M_p - W)$$

where each

$$M_i - W = V_i - (\Sigma(V_i) \cup W)$$

is a manifold which has only finitely many (path-)-components by A.4.

It follows that the union $V - W$ has only finitely many path-components. This completes the proof of Theorem 2.4.

---

* It is essential for this proof that we are working over the real numbers.

REMARK 1. Sharper estimates of the connectivity of an algebraic set have been given by THOM, "L'homologie des variétés algebriques réelles" and MILNOR, "On the Betti numbers of real varieties."

REMARK 2. It seems natural to conjecture that every difference $V - W$ actually has the homotopy type of a finite complex.

APPENDIX B

THE MULTIPLICITY OF AN ISOLATED SOLUTION OF ANALYTIC EQUATIONS

Given analytic functions $g_1, \ldots, g_m$ of m complex variables with an isolated common zero at $\mathbf{z}^0$, we have defined the *multiplicity* μ to be the degree of the associated mapping

$$\mathbf{z} \mapsto \mathbf{g}(\mathbf{z})/\|\mathbf{g}(\mathbf{z})\|$$

from the ε-sphere centered at $\mathbf{z}^0$ to the unit sphere. (Compare §7.) This appendix will justify this definition by verifying several elementary properties. First the following.

LEMMA B.1. *If the Jacobian* $(\partial g_j/\partial z_k)$ *is non-singular at* $\mathbf{z}^0$ *then* $\mu = 1$.

Proof: Consider the Taylor expansion with remainder:

$$\mathbf{g}(\mathbf{z}) = \mathbf{L}(\mathbf{z}-\mathbf{z}^0) + \mathbf{r}(\mathbf{z}) \ ,$$

where the linear transformation $\mathbf{L}$ is non-singular by hypothesis, and where

$$\|\mathbf{r}(\mathbf{z})\|/\|\mathbf{z}-\mathbf{z}^0\|$$

tends to zero as $\mathbf{z} \to \mathbf{z}^0$. Choose ε small enough so that

$$\|\mathbf{r}(\mathbf{z})\| < \|\mathbf{L}(\mathbf{z}-\mathbf{z}^0)\|$$

whenever $\|\mathbf{z}-\mathbf{z}^0\| = \varepsilon$. Then the one-parameter family of mappings

$$h_t(z) = (L(z-z^0) + tr(z))/\|L(z-z^0) + tr(z)\|, \ 0 \le t \le 1,$$

from $S_\varepsilon(z^0)$ to the unit sphere demonstrates that the degree μ of h_1 is equal to the degree of the mapping $L/\|L\|$ on $S_\varepsilon(z^0)$.

Note. The fact that the degree of $(L+r)/\|L+r\|$ on S_ε is equal to the degree of $L/\|L\|$ whenever $\|r\| < \|L\|$ throughout S_ε will be used frequently. We will refer to this fact as "Rouché'a Principle."

Now deform L continuously to the identity within the group $GL(m, C)$ consisting of all non-singular linear transformations. This is possible since the Lie group $GL(m, C)$ is connected. It follows easily that the degree of the mapping $L/\|L\|$ on $S_\varepsilon(z^0)$ is $+1$. This completes the proof.

Next consider a compact region D with smooth boundary in C^m. Assume that g has only finitely many zeros in D, and no zeros on the boundary.

LEMMA B.2. *The number of zeros of* g *within* D, *each counted with its appropriate multiplicity, is equal to the degree of the mapping*

$$z \mapsto g(z)/\|g(z)\|$$

from ∂D *to the unit sphere.*

(REMARK. For functions of one complex variable this statement is called the "principle of the argument." See for example HILLE, §9.2.2.)

Proof: Remove a small open disk about each zero of g from the region D. Then the function $g/\|g\|$ is defined and continuous throughout the remaining region D_0. Since ∂D is homologous to the sum of the small boundary spheres within D_0, it follows that the degree of $g/\|g\|$ on ∂D is equal to the sum, $\Sigma\,\mu$, of the degrees on the small spheres. [Compare MILNOR, "Topology from the differentiable viewpoint," p. 28, 36.] This completes the proof.

Again let z^0 be an isolated zero of g with multiplicity μ.

LEMMA B.3. *If* D_ε *is a disk about* $\mathbf{z}^0$ *containing no other zeros of* $\mathbf{g}$ *then, for almost all points* $\mathbf{a} \in C^m$ *sufficiently close to the origin, the equation* $\mathbf{g}(\mathbf{z}) = \mathbf{a}$ *has precisely* μ *solutions* $\mathbf{z}$ *within* D_ε.

In particular this certainly implies:

COROLLARY B.4. *The inequality* $\mu \geq 0$ *is always satisfied.*

Proof of B.3. According to the theorem of SARD, almost every point $\mathbf{a}$ of C^m is a regular value of the differentiable mapping

$$\mathbf{g} : C^m \to C^m .$$

(Compare DE RHAM, p. 10.) In other words for all $\mathbf{a}$ not belonging to some set of Lebesgue measure zero, the matrix $(\partial g_j/\partial z_k)$ is non-singular at every point $\mathbf{z}$ in the inverse image $\mathbf{g}^{-1}(\mathbf{a})$.

Given any such regular value $\mathbf{a}$, note that the solutions $\mathbf{z}$ of the system of analytic equations $\mathbf{g}(\mathbf{z})-\mathbf{a} = 0$ are all isolated, with multiplicity $+1$. (Lemma B.1.)

Choose any regular value $\mathbf{a}$ of $\mathbf{g}$ which is close enough to the origin so that

$$\|\mathbf{a}\| < \|\mathbf{g}(\mathbf{z})\|$$

for all $\mathbf{z} \in \partial D_\varepsilon$. Then according to Lemma B.2 the number of solutions of the equation $\mathbf{g}(\mathbf{z}) - \mathbf{a} = 0$ within D_ε is equal to the degree of the map $(\mathbf{g}-\mathbf{a})/\|\mathbf{g}-\mathbf{a}\|$ on ∂D_ε. (Each solution must be counted with a certain multiplicity, but we have just seen that these multiplicities are all $+1$.)

By Rouché's Principle the degree of this mapping $(\mathbf{g}-\mathbf{a})/\|\mathbf{g}-\mathbf{a}\|$ is equal to the degree μ of $\mathbf{g}/\|\mathbf{g}\|$. This completes the proof of B.3.

REMARK. It is perhaps worth interpreting these lemmas in the special case $g_j(\mathbf{z}) = \partial f/\partial z_j$ of §7. Lemma B.1 says that in the case of a *non-degenerate* critical point of f, where the Hessian matrix $(\partial^2 f/\partial z_j \partial z_k)$ is non-singular, the integer μ is $+1$. Lemma B.3 says that if we perturb f, by subtracting almost any "small" linear polynomial $a_1 z_1 + \cdots + a_m z_m$

from it, then the isolated critical point $\mathbf{z}^0$ will split up into a cluster of μ nearby critical points, all non-degenerate.

Now we are ready to prove Theorem 7.1.

THEOREM. *The multiplicity* μ *of an isolated solution of* m *polynomial equations in* m *variables is always a positive integer.*

Proof: Given a disk D_ε about $\mathbf{z}^0$ containing no other zeros of $\mathbf{g}$, choose a number η which is small enough so that

$$|\eta| < \|\mathbf{g}(\mathbf{z})\|/\varepsilon$$

for all $\mathbf{z} \in \partial D_\varepsilon$, and which is distinct from all eigenvalues of the matrix $(\partial g_j(\mathbf{z}^0)/\partial z_k)$. Then the perturbed function

$$\mathbf{g}'(\mathbf{z}) = \mathbf{g}(\mathbf{z}) - \eta(\mathbf{z}-\mathbf{z}^0)$$

has a zero of multiplicity $+1$ at $\mathbf{z}^0$, since the matrix

$$(\partial g_j'/\partial z_k) = (\partial g_j/\partial z_k - \eta\delta_{jk})$$

is non-singular at $\mathbf{z}^0$. Therefore, assuming that $\mathbf{g}'$ has only finitely many zeros within D_ε, the algebraic number $\Sigma\,\mu'$ of zeros of $\mathbf{g}'$ within D_ε is certainly ≥ 1. (All summands being ≥ 0 by B.4.) This sum is equal to the degree of $\mathbf{g}'/\|\mathbf{g}'\|$ on ∂D_ε, which is equal to the degree μ of $\mathbf{g}/\|\mathbf{g}\|$ on ∂D_ε by Rouché's Principle. Hence $\mu \geq 1$.

There remains, theoretically at least, the possibility that $\mathbf{g}'$ has infinitely many zeros within D_ε (Compare Problem 1 below.) But in that case we could subtract a small constant vector $\mathbf{a}$ from $\mathbf{g}'$, where $\mathbf{a}$ is a regular value of $\mathbf{g}'$. (Compare B.3.) Then the zeros of $\mathbf{g}'-\mathbf{a}$ are isolated, and hence there are only finitely many zeros of $\mathbf{g}'-\mathbf{a}$ within D_ε. To guarantee that $\mathbf{g}'-\mathbf{a}$ has at least one zero, we use the Inverse Function Theorem to choose a neighborhood U of $\mathbf{z}^0$ in D_ε so that $\mathbf{g}'$ maps U diffeomorphically onto an open neighborhood of the origin. Choosing $\mathbf{a}$ within $\mathbf{g}'(U)$ the equation $\mathbf{g}'(\mathbf{z}) - \mathbf{a} = 0$ certainly has a solution $\mathbf{z}$ within $U \subset D_\varepsilon$. This completes the proof that $\mu \geq 1$.

To conclude this discussion here are three problems for the reader. The first two can be established using the methods above, but the third is more difficult.

Problem 1. If $\mathbf{g}$ has no zeros on ∂D, show that it has only finitely many zeros within D.

Problem 2. If the matrix $(\partial g_j/\partial z_k)$ is singular at $\mathbf{z}^0$, show that $\mu \geq 2$.

Problem 3. The ring $C[[\mathbf{z}-\mathbf{z}^0]]$ of formal power series in the variables $z_j - z_j^0$ can be considered as a module over the subring $C[[g_1, \ldots, g_m]]$. This module is free of rank μ. Hence, if I denotes the ideal spanned by $g_1, \ldots, g_m$ in $C[[\mathbf{z}-\mathbf{z}^0]]$, then the quotient ring $C[[\mathbf{z}-\mathbf{z}^0]]/I$ has dimension μ over C. (I am told that these statements can be proved by showing first that the map $\mathbf{g}: C^m \to C^m$ induces a proper and flat map from a small neighborhood U of the origin to a small neighborhood V of the origin; and then that the direct image under g of the sheaf $\mathcal{O}_U$ of germs of holomorphic functions on U is locally free over the corresponding sheaf $\mathcal{O}_V$.)

BIBLIOGRAPHY

J. W. Alexander and G. B. Briggs, On types of knotted curves, *Annals of Math.*, *28* (1927), 562-586.

J. W. Alexander, Topological invariants of knots and links, *Trans. Amer. Math. Soc.*, *30* (1928), 275-306.

P. Alexandroff and H. Hopf, *Topologie*, Springer, 1935.

A. Andreotti and T. Frankel, The Lefschetz theorem on hyperplane sections, *Annals of Math.*, *69* (1959), 713-717.

K. Brauner, Zur Geometrie der Funktionen zweier komplexen Veränderlichen III, IV, *Abh. Math. Sem. Hamburg*, *6* (1928), 8-54.

E. Brieskorn Examples of singular normal complex spaces which are topological manifolds, *Proc. Nat. Acad. Sci. U.S.A.*, *55* (1966), 1395-1397.

______, Beispiele zur Differentialtopologie von Singularitäten, *Inventiones Math.*, *2* (1966), 1-14.

W. Browder and J. Levine, Fibering manifolds over a circle, *Comment. Math. Helv.*, *40* (1965-66), 153-160.

F. Bruhat and H. Cartan, Sur la structure des sous-ensembles analytiques réels, *C. R. Acad. Sci. Paris*, *244* (1957), 988-990.

W. Burau, Kennzeichnung der Schlauchknoten, *Abh. Math. Sem. Hamburg*, *9* (1932), 125-133.

______, Kennzeichnung der Schlauchverkettungen, *Abh. Math. Sem. Hamburg*, *10* (1934), 285-397.

H. Cartan, Quotient d'un espace analytique par un groupe d'automorphismes, *Algebraic Geometry and Topology* (Lefschetz symposium volume), Princeton Univ. Press 1957, 90-102.

C. Chevalley, Introduction to the theory of algebraic functions of one variable, *Amer. Math. Soc. Surveys #6,* 1951.

R. H. Crowell, Corresponding group and module sequences, *Nagoya Math. J., 19* (1961), 27-40.

______, The group G'/G'' of a knot group G, *Duke Math. J., 30* (1963), 349-354.

______ and R. H. Fox, *Introduction to Knot Theory,* Ginn, 1963.

P. Du Val, On isolated singularities of surfaces which do not affect the conditions of adjunction, *Proc. Cambridge Phil. Soc., 30* (1934), 453-459.

C. Ehresmann, Sur les espaces fibrés differentiables, *Compt. Rend. Acad. Sci. Paris, 224* (1947), 1611-1612.

I. Fáry, Cohomologie des variétés algébriques, *Annals of Math., 65* (1957), 21-73.

R. H. Fox, Free differential calculus, II. The isomorphism problem, *Annals of Math., 59* (1954), 196-210.

L. M. Graves, *The Theory of Functions of Real Variables,* McGraw-Hill, 1956.

R. Gunning and H. Rossi, *Analytic Functions of Several Complex Variables,* Prentice-Hall, 1965.

G.-H. Halphen, Étude sur les points singuliers des courbes algébriques planes, *Oeuvres, Tome 4,* 1-93.

E. Hille, *Analytic Function Theory, vol. 1,* Ginn, 1959.

F. Hirzebruch, The topology of normal singularities of an algebraic surface (d'après Mumford), *Séminaire Bourbaki, 15e année,* 1962/63, No. 250.

F. Hirzebruch, Singularities and exotic spheres, *Séminaire Bourbaki, 19e année*, 1966/67, No. 314.

______, O(n)-Mannigfaltigkeiten, exotische Sphären, kuriose Involutionen, (preliminary draft), March 1966.

______, and K. H. Mayer, O(n)-Mannigfaltigkeiten, exotische Sphären und Singularitäten, *Springer Lecture Notes in Mathematics, 57* (1968), 132 pages.

W. V. D. Hodge and D. Pedoe, *Methods of Algebraic Geometry, vol. 2,* Cambridge U. Press, 1952.

L. Hörmander, On the division of distributions by polynomials, *Ark. Mat. 3,* (1958), 555-568.

S. T. Hu, *Theory of Retracts,* Wayne State Univ. Press, 1965.

W. E. Jenner, *Rudiments of Algebraic Geometry,* Oxford U. Press, 1963.

K. Kähler, Über die Verzweigung einer algebraischen Funktion zweier Veränderlichen in der Umgebung einen singulären Stelle, *Math. Zeit.*, *30* (1929), 188-204.

M. Kervaire, A manifold which does not admit any differentiable structure, *Commentarii Math. Helv.*, *34* (1960), 257-270.

______ and J. Milnor, Groups of homotopy spheres I, *Annals of Math.*, *77* (1963), 504-537.

F. Klein, *Lectures on the icosahedron and the solution of equations of the fifth degree,* Dover 1956.

S. Kobayashi, Fixed points of isometries, *Nagoya Math. J.*, *13* (1958), 63-68.

S. Lang, *Introduction to Algebraic Geometry,* Interscience, 1958.

______, *Introduction to Differentiable Manifolds,* Interscience, 1962.

______, *Algebra,* Addison-Wesley, 1965.

S. Lefschetz, *Algebraic Geometry,* Princeton Univ. Press, 1953.

S. Lefschetz, *Topology* (2nd ed.), Chelsea, 1956.

J. Levine, Polynomial invariants of knots of codimension two, *Annals of Math.*, *84* (1966), 537-554.

S. Lojasiewicz, Sur le problème de la division, *Rozprawy Mat.*, *22* (1961), 57 pp., or *Studia Math.*, *18* (1959), 87-136.

______, Triangulation of semi-analytic sets, *Annali Scu. Norm. Sup. Pisa, Sc. Fis. Mat. Ser. 3, v. 18,* fasc. 4 (1964), 449-474.

J. Milnor, Construction of universal bundles II, *Annals of Math.*, *63* (1956), 430-436.

______, Morse Theory, *Annals Study #51,* Princeton Univ. Press, 1963.

______, On the Betti numbers of real varieties, *Proc. Amer. Math. Soc.*, *15* (1964), 275-280.

______, *Topology from the Differentiable Viewpoint,* Univ. Virginia Press, 1965.

______, Infinite cyclic coverings, to appear.

M. Morse, The calculus of variations in the large, *Amer. Math. Soc. Colloq. Publ. 18,* (1934).

D. Mumford, The topology of normal singularities of an algebraic surface and a criterion for simplicity, *Publ. math. No 9 l'Inst. des hautes études sci.*, Paris 1961.

L. Neuwirth, The algebraic determination of the genus of knots, *Amer. J. Math.*, *82* (1962), 791-798.

______, On Stallings fibrations, *Proc. Amer. Math. Soc.*, *14* (1963), 380-381.

F. Pham, Formules de Picard-Lefschetz généralisées et ramification des intégrales, *Bull. Soc. Math. France, 93* (1965), 333-367.

E. S. Rapaport, On the commutator subgroup of a knot group, *Annals of Math.*, *71* (1960), 157-162.

J. E. Reeve, A summary of results in the topological classification of plane algebroid singularities, *Rendiconti Sem. Mat. Torino, 14* (1954-55), 159-187.

G. de Rham, *Varietes Différentiables*, Hermann, 1955.

J. F. Ritt, Differential equations from the algebraic standpoint, *Amer. Math. Soc. Colloq. Publ., 14,* New York, 1932. (See p. 91.)

A. Sard, The measure of the critical values of differentiable maps, *Bull. Amer. Math. Soc., 48* (1942), 883-897.

J. P. Serre, *Groupes algébriques et corps de classes,* Hermann, Paris, 1959.

S. Smale, Generalized Poincaré's conjecture in dimensions greater than four, *Annals of Math., 74* (1961), 391-406.

______, On the structure of 5-manifolds, *Annals of Math., 75* (1962), 38-46.

E. Spanier, *Algebraic Topology,* McGraw-Hill, 1966.

G. Springer, *Introduction to Riemann Surfaces,* Addison-Wesley, 1957.

J. Stallings, Polyhedral homotopy spheres, *Bull. Amer. Math. Soc., 66* (1960), 485-488.

______, The piecewise linear structure of euclidean space, *Proc. Cambr. Phil. Soc., 58* (1962), 481-488.

______, On fibering certain 3-manifolds, *Topology of 3-manifolds and Related Topics,* (M. K. Fort Jr., ed.) Prentice-Hall, 1962, 95-100.

N. Steenrod, *The Topology of Fibre Bundles,* Princeton Univ. Press, 1951.

T. E. Stewart, On groups of diffeomorphisms, *Proc. Amer. Math. Soc., 11* (1960), 559-563.

R. Thom, Sur l'homologie des variétés algébriques réeles,, *Differential and Combinatorial Topology* (Morse symposium, S. Cairns ed.), Princeton Univ. Press, 1965, 255-265.

B. L. van der Waerden, Zur algebraische Geometric III; Über irreduzible algebraische Mannigfaltigkeiten, *Math. Annalen, 108* (1933), 694-698.

B. L. van der Waerden, *Einführung in die algebraische Geometrie*, Springer, 1939 (also Dover 1945).

______, *Modern Algebra*, Ungar, 1950.

C. T. C. Wall, Classification of (n − 1)-connected 2n-manifolds, *Annals of Math.*, *75* (1962), 163-198.

A. H. Wallace, *Homology Theory of Algebraic Varieties*, Pergamon Press, 1958.

______, Algebraic approximation of curves, *Canad. J. Math.*, *10* (1958), 242-278.

H. C. Wang, The homology groups of the fibre bundles over a sphere, *Duke Math. J.*, *16* (1949), 33-38.

A. Weil, Numbers of solutions of equations in finite fields, *Bull. Amer. Math. Soc.*, *55* (1949), 497-508.

H. Wendt, Die gordische Auflösung von Knoten, *Math. Zeitschr.*, *42* (1937), 680-696.

O. Zariski, On the topology of algebroid singularities, *Amer. J. Math.*, *54* (1932), 453-465.

H. Zassenhaus, *The Theory of Groups*, Chelsea, 1958.